■ The Assignment of the Absolute Configuration by NMR Using Chiral Derivatizing Agents

# The Assignment of the Absolute Configuration by NMR Using Chiral Derivatizing Agents

*A Practical Guide*

José M. Seco,
Emilio Quiñoá

AND

Ricardo Riguera

# OXFORD
UNIVERSITY PRESS

Oxford University Press is a department of the University of
Oxford. It furthers the University's objective of excellence in research,
scholarship, and education by publishing worldwide.

Oxford   New York
Auckland   Cape Town   Dar es Salaam   Hong Kong   Karachi
Kuala Lumpur   Madrid   Melbourne   Mexico City   Nairobi
New Delhi   Shanghai   Taipei   Toronto

With offices in
Argentina   Austria   Brazil   Chile   Czech Republic   France   Greece
Guatemala   Hungary   Italy   Japan   Poland   Portugal   Singapore
South Korea   Switzerland   Thailand   Turkey   Ukraine   Vietnam

Oxford is a registered trademark of Oxford University Press
in the UK and certain other countries.

Published in the United States of America by
Oxford University Press
198 Madison Avenue, New York, NY 10016

© Oxford University Press 2015

All rights reserved. No part of this publication may be reproduced, stored in
a retrieval system, or transmitted, in any form or by any means, without the prior
permission in writing of Oxford University Press, or as expressly permitted by law,
by license, or under terms agreed with the appropriate reproduction rights organization.
Inquiries concerning reproduction outside the scope of the above should be sent to the
Rights Department, Oxford University Press, at the address above.

You must not circulate this work in any other form
and you must impose this same condition on any acquirer.

Library of Congress Cataloging-in-Publication Data
Seco, J. M. (José Manuel)
The assignment of the absolute configuration by NMR using chiral derivatizing agents : a practical
guide / José M. Seco, Emilio Quiñoá, and Ricardo Riguera.
pages cm
Includes bibliographical references and index.
ISBN 978–0–19–999680–3 (alk. paper)   1. Nuclear magnetic resonance
spectroscopy.   2. Spectrum analysis.   3. Nuclear magnetic resonance spectroscopy—
Problems, exercises, etc.   4. Spectrum analysis—Problems, exercises, etc.   I. Quiñoá, E.
(Emilio)   II. Riguera, R. (Ricardo)   III. Title.
QD96.N8S43 2015
543'.66—dc23
2014042823

9 8 7 6 5 4 3 2 1
Printed in the United States of America
on acid-free paper

# CONTENTS

| | | |
|---|---|---|
| | *Preface* | *xi* |
| | *Introduction* | *xiii* |
| 1 | The Theoretical Basis for Assignment by NMR | 1 |
| | 1.1. Distinguishing Enantiomers by NMR: The Use of CDAs | 1 |
| | 1.2. Structural Characteristics of the Auxiliaries and the Substrates | 3 |
| | 1.3. Importance of the Conformation | 6 |
| | 1.4. Importance of the Aromatic Shielding Effect | 10 |
| | 1.5. Use of $^{13}$C NMR for Assignment | 12 |
| | 1.6. Simplified Approaches to Assignment by NMR | 13 |
| | 1.6.1. Single-Derivatization Method: Manipulating the Conformational Equilibrium by Temperature | 13 |
| | 1.6.2. Single-Derivatization Method: Manipulating the Conformational Equilibrium by Complexation | 14 |
| | 1.6.3. Single-Derivatization Method Based on Esterification Shifts | 16 |
| | 1.6.4. Mix-and-Shake Method: Assignment Using Resin-Bound CDAs | 18 |
| | 1.7. General Criteria for the Correct Application of the NMR Methodology | 18 |
| | 1.8. Correlation Models for the Assignment of Polyfunctional Compounds | 21 |
| | 1.9. Summary | 22 |
| 2 | Practical Aspects of the Preparation of the Derivatives | 27 |
| | 2.1. Instrumentation, Concentration, Solvent, and Temperature of the NMR Experiment | 27 |
| | 2.2. Source and Preparation of the CDAs | 27 |
| | 2.3. Preparation of the CDA Esters, Thioesters, and Amides | 28 |
| | 2.3.1. Derivatization of Alcohols, Thiols, and Cyanohydrins Using the CDA Acid | 28 |
| | 2.3.2. Derivatization of Amines Using the CDA Acid | 28 |
| | 2.3.3. Preparation of the CDA Acid Chlorides | 29 |
| | 2.3.4. Derivatization of Alcohols, Thiols, and Cyanohydrins Using the CDA Acid Chloride | 29 |
| | 2.4. Resin-Bound CDA Derivatives (Mix-and-Shake Method) | 29 |
| | 2.4.1. Preparation of Resin-Bound CDA Derivatives | 30 |
| | 2.4.2. Preparation of Acid Chloride Resins | 30 |

| | |
|---|---:|
| 2.4.3. Preparation of CDA-Resins | 31 |
| 2.4.4. Determination of the Loading of the CDA-Resins | 31 |
| 2.4.5. In-Tube Derivatization of Amines | 31 |
| 2.4.6. In-Tube Derivatization of Primary and Secondary Alcohols, Cyanohydrins, and Secondary Thiols | 32 |
| 2.4.7. In-Tube Derivatization of Amino Alcohols | 34 |
| 2.4.8. In-Tube Derivatization of Diols and Triols | 35 |
| 2.4.9. In-Tube Derivatization for Single-Derivatization Procedures | 35 |

3  Assignment of the Absolute Configuration of Monofunctional Compounds by Double Derivatization ............ 37

| | |
|---|---:|
| 3.1. Secondary Alcohols | 37 |
| 3.1.1. MPA and 9-AMA as CDAs for Secondary Alcohols | 37 |
| 3.1.2. Example 1: Assignment of the Absolute Configuration of Diacetone D-Glucose Using MPA | 39 |
| 3.1.3. Example 2: Assignment of the Absolute Configuration of (−)-Isopulegol Using 9-AMA | 40 |
| 3.1.4. Example 3: Assignment of the Absolute Configuration of (R)-Butan-2-ol Using MPA and $^{13}$C-NMR | 42 |
| 3.1.5. Simultaneous Derivatization of the Substrate with the (R)- and (S)-CDAs | 43 |
| 3.1.6. Example 4: Assignment of the Absolute Configuration of (S)-Butan-2-ol Using a 1:2 Mixture of (R)- and (S)-MPA | 44 |
| 3.1.7. Example 5: Assignment of the Absolute Configuration of (−)-Menthol Using a 2:1 Mixture of (R)- and (S)-9-AMA and $^{13}$C NMR | 45 |
| 3.1.8. MTPA as the CDA for Secondary Alcohols | 47 |
| 3.1.9. Example 6: Assignment of the Absolute Configuration of (−)-Borneol Using MTPA | 47 |
| 3.1.10. Summary | 48 |
| 3.2. β-Chiral Primary Alcohols | 51 |
| 3.2.1. Assignment of β-Chiral Primary Alcohols as 9-AMA Esters | 51 |
| 3.2.2. Example 7: Assignment of the Absolute Configuration of (S)-2-Methylbutan-1-ol Using 9-AMA | 52 |
| 3.2.3. Absolute Configuration of Primary Alcohols with Polar Groups as 9-AMA Esters | 53 |
| 3.2.4. Example 8: Assignment of the Absolute Configuration of (S)-2-Chloropropan-1-ol Using 9-AMA | 55 |
| 3.2.5. Summary | 56 |
| 3.3. Aldehyde Cyanohydrins | 58 |
| 3.3.1. Assignment of Aldehyde Cyanohydrins as MPA Esters | 58 |
| 3.3.2. Example 9: Assignment of the Absolute Configuration of (R)-2-Hydroxy-3-Methylbutanenitrile Using MPA | 59 |
| 3.3.3. Example 10: Assignment of the Absolute Configuration of (R)-2-Hydroxy-2-(4-Methoxyphenyl)Acetonitrile Using MPA and $^{13}$C NMR | 60 |

| | |
|---|---|
| 3.3.4. Summary | 63 |
| 3.4. Ketone Cyanohydrins | 63 |
| 3.4.1. Assignment of Ketone Cyanohydrins as MPA Esters | 65 |
| 3.4.2. Example 11: Assignment of the Absolute Configuration of (1$R$, 2$S$, 5$R$)-1-Hydroxy-2-Isopropyl-5-Methylcyclohexanecarbonitrile Using MPA | 65 |
| 3.4.3. Example 12: Assignment of the Absolute Configuration of ($S$)-2-Hydroxy-2,4-Dimethylpentanenitrile Using MPA and $^{13}$C NMR | 67 |
| 3.4.4. Summary | 69 |
| 3.5. Secondary Thiols | 69 |
| 3.5.1. MPA and 2-NTBA Thioesters of Secondary Thiols | 69 |
| 3.5.2. Example 13: Assignment of the Absolute Configuration of ($S$)-Butane-2-Thiol Using MPA | 71 |
| 3.5.3. Example 14: Assignment of the Absolute Configuration of ($R$)-Ethyl 2-Mercaptopropanoate Using 2-NTBA | 72 |
| 3.5.4. Example 15: Assignment of the Absolute Configuration of ($R$)-Ethyl 2-Mercaptopropanoate Using 2-NTBA and $^{13}$C NMR | 73 |
| 3.5.5. Summary | 76 |
| 3.6. α-Chiral Primary Amines | 76 |
| 3.6.1. BPG as the CDA for α-Chiral Primary Amines | 76 |
| 3.6.2. Example 16: Assignment of the Absolute Configuration of (−)-Isopinocampheylamine Using BPG | 78 |
| 3.6.3. Example 17: Assignment of the Absolute Configuration of ($S$)-Butan-2-Amine Using BPG and $^{13}$C NMR | 79 |
| 3.6.4. MPA as the CDA for α-Chiral Primary Amines | 80 |
| 3.6.5. Example 18: Assignment of the Absolute Configuration of (−)-Bornylamine Using MPA | 82 |
| 3.6.6. MTPA as the CDA for α-Chiral Primary Amines | 83 |
| 3.6.7. Example 19: Assignment of the Absolute Configuration of (−)-Bornylamine Using MTPA | 83 |
| 3.6.8. Summary | 84 |
| 3.7. α-Chiral Carboxylic Acids | 87 |
| 3.7.1. 9-AHA Esters of Carboxylic Acids | 89 |
| 3.7.2. Example 20: Assignment of the Absolute Configuration of ($S$)-3-(Acetylthio)-2-Methylpropanoic Acid Using 9-AHA | 90 |
| 3.7.3. Summary | 91 |
| **4** Assignment of the Absolute Configuration of Monofunctional Compounds by Single Derivatization | 93 |
| 4.1. Low-Temperature NMR Procedure for Secondary Alcohols | 93 |
| 4.1.1. Example 21: Assignment of the Absolute Configuration of Diacetone D-Glucose Using ($R$)-MPA | 95 |
| 4.1.2. Example 22: Assignment of the Absolute Configuration of ($R$)-Butan-2-ol Using ($S$)-MPA | 98 |

| | | |
|---|---|---|
| | 4.2. Complexation with $Ba^{2+}$: MPA Esters of Secondary Alcohols | 99 |
| | 4.2.1. Example 23: Assignment of the Absolute Configuration of (R)-Pentan-2-ol Using (S)-MPA | 100 |
| | 4.2.2. Example 24: Assignment of the Absolute Configuration of (R)-Pentan-2-ol Using (R)-MPA | 101 |
| | 4.3. Complexation with $Ba^{2+}$: MPA Amides of α-Chiral Primary Amines | 102 |
| | 4.3.1. Example 25: Assignment of the Absolute Configuration of (−)-Isopinocampheylamine Using (R)-MPA | 105 |
| | 4.4. Esterification Shifts | 105 |
| | 4.4.1. Example 26: Assignment of the Absolute Configuration of (1R, 4S)-Hydroxycyclopent-2-en-1-yl Acetate as (R)-9-AMA Ester | 107 |
| | 4.4.2. Example 27: Assignment of the Absolute Configuration of (1R, 4S)-Hydroxycyclopent-2-en-1-yl Acetate as (S)-9-AMA Ester | 108 |
| | 4.5. Summary | 109 |
| 5 | Assignment of the Absolute Configuration of Polyfunctional Compounds | 111 |
| | 5.1. *Sec/Sec*-1,2- and *Sec/Sec*-1,n-Diols | 111 |
| | 5.1.1. Double-Derivatization Methods: MPA, 9-AMA, and MTPA | 111 |
| | 5.1.2. Example 28: Assignment of the Absolute Configuration of Heptane-2,3-Diol (*Syn*) | 116 |
| | 5.1.3. Example 29: Assignment of the Absolute Configuration of Heptane-2,3-Diol (*Anti*) | 118 |
| | 5.1.4. Example 30: Assignment of the Absolute Configuration of 1,4-Bis-O-(4-Chlorobenzyloxy)-D-Threitol (*Syn*) Using $^{13}C$ NMR | 119 |
| | 5.1.5. Single-Derivatization Methods: MPA | 120 |
| | 5.1.6. Example 31: Assignment of the Absolute Configuration of a Pure Isomer of 3,4-Dihydroxy-5-Methylhexan-2-One | 124 |
| | 5.1.7. Example 32: Assignment of the Absolute Configuration of Another Isomer of 3,4-Dihydroxy-5-Methylhexan-2-One | 124 |
| | 5.1.8. Summary | 125 |
| | 5.2. *Sec/Sec*-1,2-Amino Alcohols | 127 |
| | 5.2.1. Double-Derivatization Method: MPA | 127 |
| | 5.2.2. Example 33: Assignment of the Absolute Configuration of 2-Aminopentan-3-ol (*Syn*) | 130 |
| | 5.2.3. Example 34: Assignment of the Absolute Configuration of Methyl 4-Amino-3-Hydroxy-5-Phenylpentanoate (*Anti*) | 132 |
| | 5.2.4. Summary | 133 |
| | 5.3. *Prim/Sec*-1,2-Diols | 134 |
| | 5.3.1. Double-Derivatization Methods: MPA | 135 |

5.3.2. Example 35: Assignment of the Absolute Configuration
 of (S)-Propane-1,2-Diol Using MPA                                    138
5.3.3. Double-Derivatization Methods: 9-AMA                           140
5.3.4. Example 36: Assignment of the Absolute Configuration
 of (S)-Propane-1,2-Diol Using 9-AMA                                  143
5.3.5. Example 37: Assignment of the Absolute Configuration
 of (R)-2,3-Dihydroxypropyl Stearate Based Only on the
 Methylene Hydrogens                                                  144
5.3.6. Example 38: Assignment of the Absolute Configuration
 of (R)-1-Phenylethane-1,2-Diol Based Only on the
 Methylene Hydrogens                                                  145
5.3.7. Single-Derivatization Method: MPA                              147
5.3.8. Example 39: Assignment of the Absolute Configuration
 of (S)-Propane-1,2-Diol                                              150
5.3.9. Example 40: Assignment of the Absolute Configuration
 of (R)-Propane-1,2-Diol                                              152
5.3.10. Summary                                                       154
5.4. *Sec/Prim*-1,2-Amino Alcohols                                    155
5.4.1. Double-Derivatization Methods: MPA and the Use of R
 and Methylene Hydrogens                                              155
5.4.2. Example 41: Assignment of the Absolute Configuration
 of (S)-2-Aminopropan-1-ol Based on R and Methylene
 Hydrogens                                                            157
5.4.3. Double-Derivatization Methods: The Use of OMe and
 CαH Signals for Assignment                                           157
5.4.4. Example 42: Assignment of the Absolute Configuration
 of (S)-2-Aminopropan-1-ol Using $\Delta\delta^{RS}$ of OMe and
 CαH signals                                                          159
5.4.5. Example 43: Assignment of the Absolute Configuration
 of (S)-2-Aminopropan-1-ol Using the Separation of
 OMe and CαH Signals                                                  161
5.4.6. Example 44: Assignment of the Absolute Configuration
 of (R)-2-Amino-3-Methylbutan-1-ol Using the
 Separation of OMe and CαH Signals                                    163
5.4.7. Single-Derivatization Method: MPA                              165
5.4.8. Example 45: Assignment of the Absolute Configuration
 of (S)-2-Aminopropan-1-ol by Low-Temperature NMR
 of a Single Derivative                                               167
5.4.9. Example 46: Assignment of the Absolute Configuration
 of (R)-2-Aminopropan-1-ol by Low-Temperature NMR
 of a Single Derivative                                               168
5.4.10. Summary                                                       171
5.5. *Prim/Sec*-1,2-Amino Alcohols                                    172
5.5.1. Double-Derivatization Methods: MPA and the Use of R
 and Methylene Hydrogens                                              173

5.5.2. Double-Derivatization Methods: The Use of OMe and
CαH Signals for Assignment ... 174
5.5.3. Example 47: Assignment of the Absolute Configuration
of (*S*)-1-Aminopropan-2-ol Based on R and Methylene
Hydrogens ... 175
5.5.4. Example 48: Assignment of the Absolute Configuration
of (*S*)-1-Aminopropan-2-ol Using $\Delta\delta^{RS}$ of OMe and
CαH Signals ... 177
5.5.5. Example 49: Assignment of the Absolute Configuration
of (*S*)-1-Aminopropan-2-ol Using the Separation of the
OMe and CαH Signals ... 178
5.5.6. Example 50: Assignment of the Absolute Configuration
of (*R*)-1-Aminoheptan-2-ol Using the Separation of the
OMe and CαH Signals ... 179
5.5.7. Single-Derivatization Method: MPA ... 181
5.5.8. Example 51: Assignment of the Absolute Configuration
of (*S*)-1-Aminopropan-2-ol by Low-Temperature NMR
of a Single Derivative ... 184
5.5.9. Example 52: Assignment of the Absolute Configuration
of (*R*)-1-Aminopropan-2-ol by Low-Temperature NMR
of a Single Derivative ... 186
5.5.10. Summary ... 187
5.6. *Prim/Sec/Sec*-1,2,3-Triols ... 188
5.6.1. Double-Derivatization Method: MPA ... 188
5.6.2. Example 53: Assignment of the Absolute Configuration
of Hexane-1,2,3-Triol (*Syn*) ... 192
5.6.3. Example 54: Assignment of the Absolute Configuration
of Hexane-1,2,3-Triol (*Anti*) ... 193
5.6.4. Summary ... 193

6　Exercises ... 197

*References* ... *225*
*Index* ... *241*

# PREFACE

This textbook is written to help students and researchers to obtain the absolute configuration of organic compounds by nuclear magnetic resonance (NMR) using arylalkoxyacetic acids as chemical derivatizing agents (CDAs). Its origin can be traced to the postgrad courses I delivered in several universities over the years and to our investigations on this topic over a fifteen-year period. Now that we have decided to move to other fields of research, it is the perfect time to put together all our experience and knowledge in a way that makes the technique available for general users.

Our contribution to this field spanned from a study of the theoretical foundations to the development of new and more efficient auxiliary reagents, greatly expanding its range of applications. Those efforts helped to transform what at the beginning was almost no more than a scientific curiosity into a methodology that allows the determination of the absolute configuration of more than a dozen different classes of compounds by using $^1$H and/or $^{13}$C NMR. Nowadays, the assignment by NMR is the easiest and cheapest method in use, and it is available to all laboratories possessing basic NMR instrumentation.

The book is intended to be primarily practical but containing the minimal theoretical background necessary to understand the foundations and the results, and therefore, it is full of NMR spectra, experimental details, and examples of applications. All this material comes from our own laboratories and has been selected with teaching as the objective. Most of the figures were originally prepared and used as slides for projection and can be very useful for specialized courses. Naturally, any error in the book can only be our responsibility.

Finally, I would like to use these lines to acknowledge my coauthors, Emilio Quiñoá and José Manuel Seco, not just because this book would not exist without their contributions, but also because their presence in the laboratory and collaboration over the years has been to me a constant source of ideas, good work, and happy moments.

Santiago de Compostela, June 2014
Ricardo Riguera

# INTRODUCTION

Over the years, the assignment of the absolute configuration of organic compounds [1–4] has been carried out using techniques such as circular dichroism (CD) [5–7] in solution, and X-ray diffraction [8–9] in the solid state. More recently, another two spectroscopic techniques have been introduced for the assignment in solution: vibrational circular dichroism (VCD) [10–12]—a technique currently available at a limited number of laboratories—and nuclear magnetic resonance (NMR) [13–15]. The latter, due to the extensive presence and use of NMR instrumentation in current research laboratories, can be considered to be of much more general applicability.

The use of NMR for the assignment of the absolute configuration of organic compounds is particularly useful in cases where the amount of the sample is limited, no monocrystals are available, or a rapid and inexpensive method is needed. In the last two decades, a great deal of effort has been invested in order to explain, to rationalize, and to extend the use of this methodology, which was initiated by Mosher and his colleagues in the 1960s [16–19]. Detailed descriptions, as well as other reviews of this topic, can be found in the literature [20–34].

The book shows in a direct way the foundations of the method, the general characteristics of arylalkoxyacetic acids as auxiliaries for chiral derivatizing agents (CDAs) [13–15], the classes of compounds whose absolute configuration can be assigned, and the scope and limitations of this technique. It is organized in such a way that Chapter 1 is devoted to the theoretical foundations, and Chapter 2 is dedicated to the practical aspects related to the NMR experiments, that is, the preparation of derivatives, reagent-supported resins, and auxiliary reagents. Chapters 3–5 discuss the application of this methodology for the assignment of configuration of different classes of organic compounds, including numerous examples and spectra. We have included in the final chapter, Chapter 6, a series of problems that can be used to test the knowledge attained by studying this book.

In order to facilitate its use for teaching, all the figures and spectra of this book in colour can be obtained from the publisher in a format appropriate for projection.

# The Assignment of the Absolute Configuration by NMR Using Chiral Derivatizing Agents

# 1 The Theoretical Basis for Assignment by NMR

## ■ 1.1. DISTINGUISHING ENANTIOMERS BY NMR: THE USE OF CDAS

The nuclear magnetic resonance (NMR) spectra of two enantiomers are identical. Thus, the first step in using NMR to distinguish between two enantiomers should be to produce different spectra that eventually can be associated with their different stereochemistry (i.e., the assignment of their absolute configuration). Therefore, it is necessary to introduce a chiral reagent in the NMR media. There are two ways to address this problem. One is to use a chiral solvent, or a chiral agent, that combines with each enantiomer of the substrate to produce diastereomeric complexes/associations that lead to different spectra. This is the so-called chiral solvating agent (CSA) approach; it will not be further discussed here [33–34].

The second approach is to use a chiral auxiliary reagent [13–15] (i.e., a chiral derivatizing agent; CDA) that bonds to the substrate by a covalent linkage. Thus, in the most general method, the two enantiomers of the auxiliary CDA react separately with the substrate, giving two diastereomeric derivatives whose spectral differences carry information that can be associated with their stereochemistry.

The CDA method that employs arylalcoxyacetic acids as auxiliaries is the most frequently used. It can be applied to a number of monofunctionals [14–15] (secondary alcohols [35–43], primary alcohols [44–46], aldehyde [47] and ketone cyanohydrins [48–49], thiols [50–51], primary amines [52–56], and carboxylic acids [57–58]), difunctional [13] (*sec/sec*-1,2-diols [59–61], *sec/sec*-1,2-amino alcohols [62], *prim/sec*-1,2-diols [63–65], *prim/sec*-1,2-aminoalcohols, and *sec/prim*-1,2-aminoalcohols [66–68]), and trifunctional (*prim/sec/sec*-1,2,3-triols [13, 69–70]) chiral compounds. Its scope and limitations are well established, and its theoretical foundations are well known, making it a reliable tool for configurational assignment.

Figure 1.1 shows a summary of the steps to be followed for the assignment of absolute configuration of a chiral compound with just one asymmetric carbon and with substituents that, for simplicity, are assumed to resonate as singlets.

Step 1 (Figure 1.1a): A substrate of unknown configuration *(?)* is separately derivatized with the two enantiomers of a chiral auxiliary reagent, (R)-Aux and (S)-Aux, producing two diastereomeric derivatives.

Step 2 (Figure 1.1b): Comparison of the $^1$H-NMR chemical shifts of the $L_1$ and $L_2$ substituents, linked to the asymmetric carbon of the substrate in the two resulting derivatives, allows the locating of $L_1$ and $L_2$ in their place around the asymmetric carbon, which is the absolute configuration.

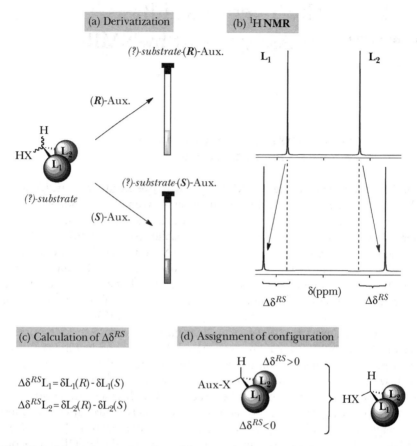

*Figure 1.1.* Schematic representation of the general procedure for the assignment of the absolute configuration by comparison of the ¹H-NMR spectra of the (R)- and the (S)-CDA derivatives of the substrate.

As can be observed, those two substituents resonate closer in one derivative and further apart in the other. It is this difference that allows locating the $L_1$ and $L_2$ substituents in the space around the asymmetric carbon.

Step 3 (Figure 1.1c): In practice, this operation is usually carried out by a subtraction defined as the chemical shifts of $L_1$ and $L_2$ in the (R) derivative minus their values in the (S) derivative. This difference is named $\Delta\delta^{RS}$ [$\Delta\delta^{RS}L=\delta L(R)-\delta L(S)$; see Figure 1.1c] and has a negative sign for one substituent and a positive sign for the other (negative for $L_1$ and positive for $L_2$, which is shown in Figure 1.1d).

Not every CDA is adequate for the assignment of any of the classes of compounds mentioned above. On the contrary, only certain CDAs have proven to give reliable assignments based on a correlation between the chemical shifts of the substituents (the sign of $\Delta\delta^{RS}$) and their spatial location.

Step 4 (Figure 1.1d): This correlation can be conveniently represented graphically as shown in Figure 1.1d, where the substituent that experimentally presents a positive $\Delta\delta^{RS}$ sign (more shielded in the (S) derivative than in the (R) derivative;

$L_2$ in the illustration of Figure 1.1) is placed in the back position of the asymmetric carbon, and the one giving the negative $\Delta\delta^{RS}$ sign (less shielded in the (S) derivative than in the (R) derivative; $L_1$ in Figure 1.1d) is placed in the front.

In Section 1.2, we will describe the general characteristics of these CDA auxiliaries, the structural classes of substrates that can be studied, and which is the best chiral auxiliary for each class of substrate; in Chapters 3–5, we will discuss the chemical shifts-stereochemistry correlation.

## ■ 1.2. STRUCTURAL CHARACTERISTICS OF THE AUXILIARIES AND THE SUBSTRATES

In the literature, a number of reagents have been proposed as CDAs, but many of these have been validated with such a small number of substrates of known absolute configuration that the evidence for their reliability is not strong [14, 20–34]. Also, in only a few cases have the theoretical bases for their function as CDAs been studied [13–14]. We will focus only on one class of CDAs, the arylmethoxyacetic acids (AMAAs) and analogues [13–14], because they are of application to a large variety of substrates, the resulting assignments have been validated with a representative series of compounds of known absolute configuration, and their mechanism of action is well known. The structural characteristic of these auxiliaries is depicted in Figure 1.2a.

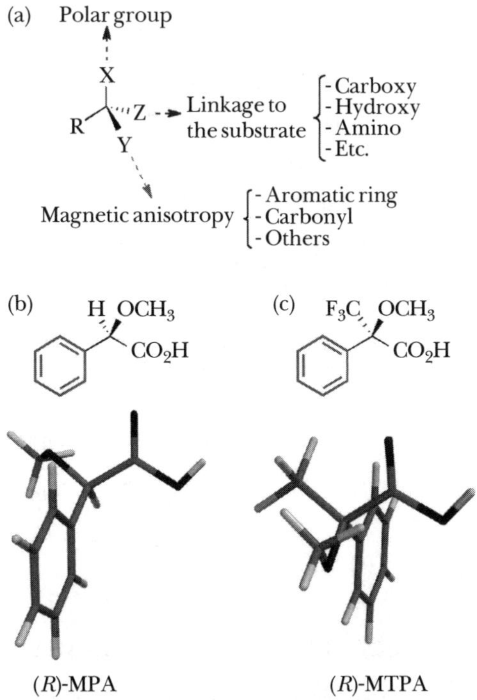

Figure 1.2. Structural characteristics of the CDA auxiliaries discussed in this book.

4 ■ The Assignment of the Absolute Configuration by NMR

All these CDAs possess a functional Z group, which is necessary for covalent bonding to the substrate; a Y group with magnetic anisotropy to selectively shield the substituents $L_1$ and $L_2$ in the NMR spectra; and an X group that should be polar in order to direct the conformational composition of the auxiliary part of the derivative.

The most common CDAs are α-methoxyphenylacetic acid (MPA, Figure 1.2b) and α-methoxytrifluoromethylphenylacetic acid (MTPA, Figure 1.2c), where Z = carboxylic acid allows their bonding to alcohols, amines, and thiols, Y = a phenyl ring, and X = methoxy in MPA or X = trifluoromethyl in MTPA. Other CDAs with the same general structure but different Z, Y, and X groups are shown in Figure 1.3. Each one has been designed for a particular application, which will be presented in the remaining sections of Chapter 1 [13–14].

Naturally, the Z groups on the auxiliary and the functionality on the substrate should be complementary in order to allow a covalent bond to be established after an experimentally simple derivatization, and thus a carboxylic acid like MPA can be used to derivatize alcohols, amines, thiols, and others, while the hydroxy group of ethyl 2-hydroxy-2-(9-anthryl) acetate (9-AHA) allows the assignment of chiral carboxylic acids as substrates.

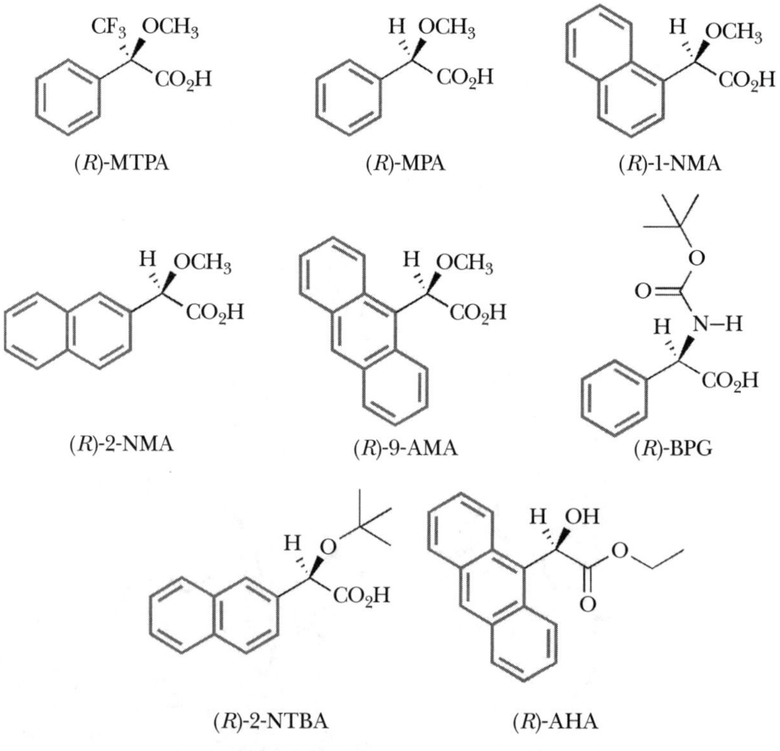

*Figure 1.3.* CDAs (arylalcoxyacetic acid types and similar) for the NMR assignment of absolute configuration of the substrates shown in Figure 1.4.

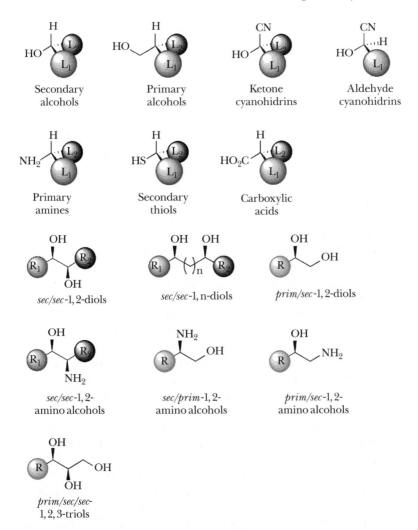

Figure 1.4. Mono- and polyfunctional compounds whose absolute configuration can be established by NMR.

The CDAs shown in Figure 1.3 have proven to produce reliable configurational assignment of the substrates listed in Figure 1.4. These comprise monofunctional compounds (in this context, monofunctional means that only one unit of the CDA reacts with the substrate, and therefore, the derivative is a monoderivative; examples include *sec*-alcohols, *prim*-alcohols, *prim*-amines, *sec*-thiols, cyanohydrines, and carboxylic acids) and bi- and trifunctional compounds (two or three units of the CDA are incorporated into the derivative; examples include *sec/sec*-1,2-diols, *sec/sec*-1,2-amino alcohols, *prim/sec*-1,2-diols, *prim/sec*-1,2-amino alcohols, *sec/prim*-1,2-amino alcohols, and *prim/sec/sec*-1,2,3-triols); see Figure 1.4.

## 1.3. IMPORTANCE OF THE CONFORMATION

Let us imagine that the substrate represented in Figure 1.1 is a secondary alcohol (X = O) and that we will use MPA as the auxiliary reagent. We must prepare the corresponding ester derivatives with the two enantiomers of the CDA, that is, the (R)-MPA and the (S)-MPA esters, and record their $^1$H-NMR spectra. Once the compounds are dissolved in the NMR tube, their conformational composition or, more precisely, the position of the phenyl ring with respect to the alcohol part of the ester, is the main factor determining the chemical shifts. Thus, conformational analysis of the MPA esters [37, 71] is necessary in order to understand and predict the chemical shifts of $L_1/L_2$ in those two derivatives.

Experimental (e.g., dynamics; low-temperature NMR; nuclear Overhauser effect, NOE; and circular dichroism, CD) and theoretical data (e.g., from molecular mechanics, and semiempirical, ab initio, and aromatic shielding effect calculations) [37, 71] have indicated that those two derivatives can be represented by two main conformers in equilibrium by rotation around the Cα-C=O bond: the synperiplanar (*sp*) and the antiperiplanar (*ap*) conformers (Figure 1.5), both having the methoxy, the carbonyl, and the C-H groups in the same plane but differing in their orientation. The methoxy and the carbonyl are *syn* in the *sp* conformer (the most populated) and *anti* in the *ap* conformer (the least populated).

Thus, in the NMR tube, we have the fast equilibrium shown in Figure 1.6 for the (R)-MPA and the (S)-MPA derivatives. The position of the phenyl ring in each conformer determines a selective shielding that mainly affects only one of the $L_1/L_2$ substituents.

In the (R)-MPA ester, the phenyl ring shields $L_1$ in the *sp* conformer, but $L_2$ in the *ap* conformer (Figure 1.6a). As the conformer *sp* is the most populated, $L_1$ is the most shielded substituent. The reverse situation happens with the (S)-MPA ester. In this case, $L_2$ is shielded in the *sp* conformer, while $L_1$ is shielded in the *ap* conformer (Figure 1.6b). As a result, at equilibrium, $L_2$ is the most heavily shielded substituent.

The equilibrium is fast in the NMR time scale; therefore, the chemical shifts observed are the mean values of those in each conformer. Since the *sp* form is more abundant than the *ap* form in MPA esters, their NMR spectra can be interpreted by assuming that the MPA ester is represented by just the *sp* conformer. In this way, a Dreiding or similar model (Figure 1.6, bottom) representing the *sp*

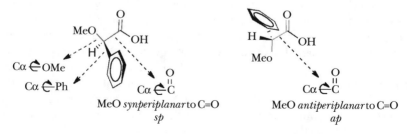

*Figure 1.5. Sp* and *ap* conformers in MPA.

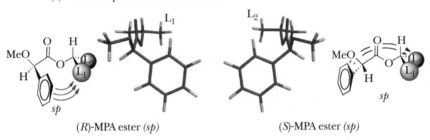

*Figure 1.6.* Conformational equilibrium in the (R)- and in the (S)-MPA derivatives of a secondary alcohol and Dreiding models of the *sp* conformer of the (R)- and the (S)-MPA esters showing the spatial orientation of $L_1/L_2$ with respect to the phenyl ring.

form can be used to explain the chemical shift changes observed in the (R)- and the (S)-MPA ester derivatives of the secondary alcohol.

Thus, the two derivatives should produce the spectra represented in Figure 1.7 (for simplicity, we assumed that $L_1/L_2$ produce singlets; the scale is arbitrary). In the (R)-MPA ester, $L_1$ is shielded in the *sp* conformer and is not affected in the *ap* conformer (Figure 1.7a), while $L_2$ is shielded in the *ap* conformer and is not affected in the *sp* conformer (Figure 1.7a). The opposite situation is observed in the (S)-MPA ester (Figure 1.7c). Of course, due to the conformational exchange, the shielding and chemical shifts observed for $L_1$ and $L_2$ are average values, and their magnitudes depend on the relative population of the *sp* and *ap* conformers. The resulting average NMR spectra are shown in Figures 1.7b and 1.7d: $L_1$ is more shielded in the (R)- than in the (S)-MPA ester, and $L_2$ is more shielded in the (S)- than in the (R)-MPA ester. Therefore, the chemical shifts of $L_1/L_2$ reflect the spatial position of those substituents with respect to the phenyl ring that selectively shields the group located on the same side.

In this way, the aromatic shielding effect produced by the phenyl ring of the auxiliary allows the translation of the stereochemical information contained in the auxiliary part (CDA, known absolute configuration) to the substrate (alcohol, unknown absolute configuration).

8 ■ The Assignment of the Absolute Configuration by NMR

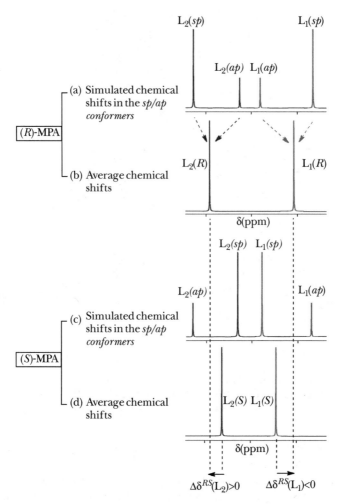

*Figure 1.7.* Conceptual representations of the $^1$H-NMR spectra of (R)-MPA and (S)-MPA esters and the resulting $\Delta\delta^{RS}$ signs and values.

This means that comparison of the chemical shifts of $L_1$ and $L_2$ in the two derivatives is enough to determine which one of the two substituents is placed on the same side as the phenyl ring in each derivative, and this is sufficient for determining the stereochemistry around the asymmetric carbon of the alcohol.

In practice, this is conveniently done by a subtraction defined as the chemical shift in the (R)-MPA derivative minus that observed for the same substituent in the (S)-MPA derivative, and this is expressed as $\Delta\delta^{RS}$. For instance, $\Delta\delta^{RS}$ $L_1 = \delta L_1$ in the (R)-MPA derivative minus $\delta L_1$ in the (S)-MPA derivative; and similarly for $L_2$.

According to this, the substituent $L_1$ in Figure 1.7 has a negative difference associated with its location in the front part of the asymmetric carbon of the alcohol (to the same side as the phenyl in the (R)-MPA ester, Figures 1.6a and 1.6c), while substituent $L_2$ presents a positive difference associated with its location at the back (Figures 1.6b and 1.6c).

This correlation between the signs of $\Delta\delta^{RS}$ for $L_1$ and $L_2$ and the position in the space of these substituents is the key to the assignment of absolute configuration. Naturally, if we change the CDA for one with a different conformational composition, we will obtain different $\Delta\delta^{RS}$ signs for $L_1$ and $L_2$ for the same absolute configuration.

This correlation between the spatial position of the $L_1/L_2$ substituents and their $\Delta\delta^{RS}$ signs is specific for each CDA-substrate couple and can therefore be used for the assignment of a configuration for that class of chiral compounds (secondary alcohols in the example).

It is important to point out that the generality of the correlation between the sign of $\Delta\delta^{RS}$ and the stereochemistry is based not only on studies of preferred conformations and aromatic shielding effect calculations [37, 39], but it has been experimentally validated with a large number of compounds of varied structure and known absolute configuration.

To illustrate the simplicity of applying this methodology, Figure 1.8 shows schematically the spectra and steps to be followed for the assignment of configuration of a pure enantiomer of pentan-2-ol, taken as a representative example. It also includes illustrative stereomodels.

According to the steps formulated in Figure 1.1, the methyl and propyl chains linked to the asymmetric carbon of pentan-2-ol constitute the $L_1/L_2$ substituents. The (R)- and (S)-MPA ester derivatives have been prepared, their spectra recorded, and the signals assigned (step 1; Figure 1.8a). Now, we focus on the signals for the two L groups (methyl and propyl) and compare the position in the spectra of the signals corresponding to their protons in the two derivatives (step 2; Figure 1.8b), observing how the methyl 1' is more shielded in the (S)-MPA ester than in the (R)-MPA ester, while all the protons at the propyl chain (methylene 3', methylene 4', and methyl 5') are more shielded in the (R)-MPA derivative than in the (S)-MPA derivative. According to step 3, the $\Delta\delta^{RS}$ differences are obtained by a subtraction defined as the chemical shifts of the protons in the (R)-MPA ester minus those in the (S)-MPA ester [$\Delta\delta^{RS}L = \delta L(R) - \delta L(S)$], leading to a positive difference of $\Delta\delta^{RS} = +0.13$ for Me(1') and negative differences for all the protons in the propyl chain [$\Delta\delta^{RS} = -0.09$ for H(3'); $-0.23$ for H(4'); $-0.14$ for Me(5'); Figure 1.8c].

A graphic stereomodel that expresses the correlation between the spatial location of the substituents $L_1/L_2$ and their $\Delta\delta^{RS}$ signs is shown in Figure 1.8d. This picture shows that, in general for the MPA esters of secondary alcohols, the substituent at the front of the tetrahedron ($L_1$) has a negative $\Delta\delta^{RS}$ sign, while the substituent with a positive $\Delta\delta^{RS}$ sign is located at the back ($L_2$). Comparison (step 4) of the experimentally obtained $\Delta\delta^{RS}$ signs for the methyl and the propyl groups of pentan-2-ol with the stereomodel leads to placing the methyl (1') at $L_2$ and the propyl chain at $L_1$. This corresponds to the (R) configuration, thus identifying the compound as (R)-pentan-2-ol.

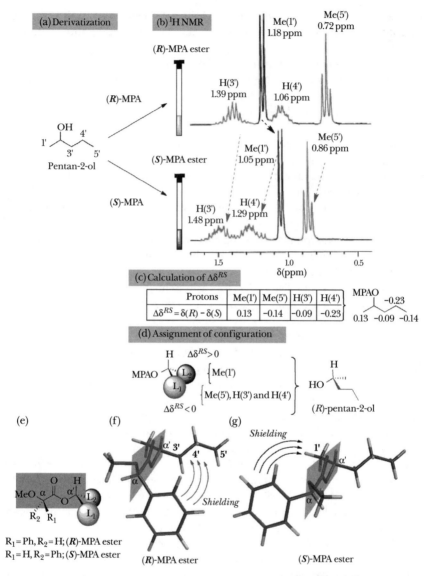

Figure 1.8. General procedure for the assignment of the absolute configuration of a secondary alcohol (pentan-2-ol) of unknown configuration. Illustrative stereomodels are also shown.

## 1.4. IMPORTANCE OF THE AROMATIC SHIELDING EFFECT

If the shifts shown in previous sections are due to the aromatic shielding effect [37, 39], then a modification of the aromatic ring current should produce higher or lower shifts on the same substrate. This has been proven by introducing donor or acceptor substituents at the phenyl ring of MPA.

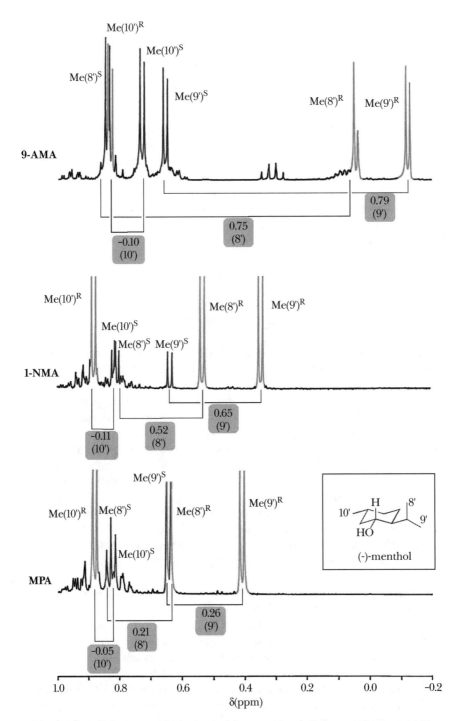

*Figure 1.9.* ¹H-NMR spectra of (−)-menthol derivatized with different AMAAs: 9-AMA, 1-NMA, and MPA.

Similarly, replacing the phenyl ring of MPA by an aromatic system (such as naphthalene or anthracene) that has a greater ring current also leads to greater $\Delta\delta^{RS}$ values [37]. An illustrative example is shown in Figure 1.9, where the spectra of (−)-menthol derivatized with different AMAAs [9-anthrylmethoxyacetic acid, 9-AMA; 1-naphthylmethoxyacetic acid, 1-NMA; and MPA] leads to larger separation of the signals and larger $\Delta\delta^{RS}$ values (e.g., with 9-AMA, it is around three times larger than with MPA).

Naturally, the shape of the π system in the diverse aromatic rings is different, and therefore, this has an effect on the direction and shape of the aromatic shielding effect. As a consequence, the zone of maximum shielding along the structures of $L_1$ and $L_2$ can change for different CDAs [39].

The relationship between the intensity and shape of the ring current and $\Delta\delta^{RS}$ has also been shown by theoretical calculations [37, 39]. This has important consequences in practice, because it allows the researcher to select the AMAA that is best suited for the particular structural characteristics of the given substrate or to select the AMAA that will increase the $\Delta\delta^{RS}$ values when they are too small and higher shifts are necessary for a good assignment [36, 39]. This will be further discussed in Chapters 3–5, below.

## ■ 1.5. USE OF $^{13}$C NMR FOR ASSIGNMENT

A second practical consequence of the above relationship is due to the use of the NMR of nuclei other than $^1$H, in particular, $^{13}$C NMR [72–80]. The aromatic shielding effect is a general phenomenon that affects not only $^1$H but also all nuclei that surround the anisotropic group. Changes in the intensity of the chemical shift generated by a carbon nucleus are much lower than on a proton [81], but in many cases, these are distinguishable and higher than the experimental error of the chemical shift measurement [72]. Therefore, the above discussion for $^1$H NMR can also be applied to $^{13}$C NMR [72]; examples in the literature have demonstrated that the variations in $\Delta\delta^{RS}$ values for the $^1$H and the $^{13}$C of a particular substrate follow exactly the same trend [72].

This means that to determine the absolute configuration of a substrate derivatized with a CDA, we are not limited to the $^1$H shifts but can also use the $^{13}$C shifts and the $\Delta\delta^{RS}$ that corresponds to them. As the phenomenon causing the shifts is the same for both nuclei (aromatic shielding effect), the $\Delta\delta^{RS}$ sign distribution of both $^{13}$C and $^1$H should be identical for the same stereochemistry, and therefore, the assignment is carried out using the very same correlation between the $\Delta\delta^{RS}$ signs and the spatial locations of $L_1/L_2$.

In fact, the NMR tubes containing the (R)- and the (S)-CDA derivatives could be used for both $^{13}$C and $^1$H-NMR spectroscopies, and the determination could be carried out by considering the $^1$H and $^{13}$C $\Delta\delta^{RS}$ signs either separately or as a single data set.

Nevertheless, one should always keep in mind that the $^{13}$C $\Delta\delta^{RS}$ values are in general very small, in some cases close to the experimental error of the shift measurement. Therefore, some $^{13}$C data may be useless for assignment, because no reliable signs can be experimentally derived from them [82].

## 1.6. SIMPLIFIED APPROACHES TO ASSIGNMENT BY NMR

As indicated in the previous sections, to make the assignment, it is necessary to prepare two derivatives (the "double-derivatization procedure") and to compare their NMR spectra [13–15]. Over the years, some efforts have been made to simplify this procedure. One approach (the "single-derivatization procedure") has the objective of reducing—from two to one—the number of derivatives that are necessary for the assignment. This has been demonstrated to work well with secondary alcohols [41–43], α-chiral primary amines [55–56], *sec/sec*-1,2-diols [61], *prim/sec*-1,2-diols [65], and *prim/sec*- and *sec/prim*-1,2-aminoalcohols [68]. It can be implemented in three different ways, all of which are related to the conformational composition of the CDA derivatives.

The second approach (the "mix-and-shake procedure") is completely general for all substrates and CDAs, and it uses resin-bound CDAs to reduce the amount of laboratory benchwork required to prepare/purify the CDA derivatives [83–84].

### 1.6.1. Single-Derivatization Method: Manipulating the Conformational Equilibrium by Temperature

The conformational equilibrium indicated in Figure 1.10a for the MPA derivatives of secondary alcohols shows that there are two main conformers—*sp* and *ap*, with the former more stable than the latter—in a fast equilibrium [36–37, 41]. Therefore, the chemical shifts observed are the mean values between those of the *sp* and the *ap* conformers and their populations.

In accordance with thermodynamic principles, if the temperature of a system in uneven equilibrium is modified [37, 41], the relative populations of the conformers should also change: a decrease in the temperature leads to an increase in the population of the most stable form and a corresponding decrease in the population of the least stable one.

Thus, the NMR spectrum of an AMA derivative [e.g., (*R*)-MPA ester, Figure 1.10] taken at a low temperature will be different from the spectrum taken at room temperature (usually 300 K); this is associated with an increase in the number of molecules in the *sp* conformer (more stable) and a decrease in the number of molecules in the *ap* conformation (less stable).

As a consequence, the contribution of the *sp* and *ap* forms to the average chemical shifts of $L_1$ and $L_2$ will be modified: the signals due to the $L_1$ substituent located on the same side as the aryl ring in the *sp* conformation (affected by the aromatic shielding, Figures 1.10a and b) will be more shielded, while the signals due to the $L_2$ substituent located on the same side as the aryl ring in the *ap* conformer (less stable), will appear less shielded at lower temperatures (Figures 1.10a and c).

This correlation between the location in space of $L_1/L_2$ and the changes in their chemical shifts when the temperature is modified allow us to assign the absolute configuration using just one derivative [either the (*R*) or the (*S*)] by comparison of its NMR spectra taken at two different temperatures [41]. This correlation is expressed for each L substituent in a form quite similar to the one used before in

14 ■ The Assignment of the Absolute Configuration by NMR

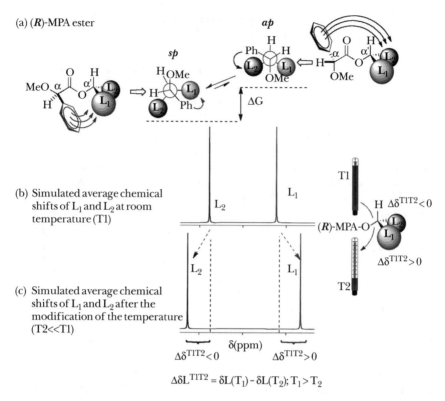

Figure 1.10. Conceptual representation of the single-derivatization procedure, based on modification of the temperature.

the double-derivatization method, that is, as the difference represented by $\Delta\delta^{T1T2}$ (L) = $\delta$(L) at a higher temperature ($T_1$) minus $\delta$(L) at a lower temperature ($T_2$) [$\Delta\delta L^{T1T2} = \delta L(T_1) - \delta L(T_2)$; $T_1 > T_2$]. In the example shown in Figure 1.10, $\Delta\delta^{T1T2}$ is positive for $L_1$ and negative for $L_2$.

Naturally, the distribution of signs for $L_1$ and $L_2$ in a particular substrate derivatized with (R)-MPA should be opposite those obtained using the (S)-MPA derivatives of the same substrate.

This approach has been shown to work well with MPA as the CDA and secondary alcohols as substrates [41]. It can be applied to *prim/sec*-1,2-diols [65] and *prim/sec*- and *sec/prim*-1,2-aminoalcohols [68], and, with a few limitations, it can be applied to *sec/sec*-diols [61]. With these types of substrates, the absolute configuration can be confidently assigned using only one derivative and taking its NMR spectra at two different temperatures.

## 1.6.2. Single-Derivatization Method: Manipulating the Conformational Equilibrium by Complexation

A different way to shift the conformational equilibrium in a controlled way without having to change the temperature of the NMR probe has been devised for the MPA derivatives of secondary alcohols [42] and α-chiral primary amines [55–56].

It consists of taking the NMR spectra of the MPA derivative (as before, just one is needed) in MeCN-$d_3$ as a solvent and at room temperature, and comparing the spectra before and after the addition of a small quantity of $Ba^{2+}$ perchlorate to the NMR tube.

In these conditions, the barium cation forms a chelate with the carbonyl and methoxy groups of the MPA. The complex formed with the *sp* conformer is more stable than that formed with the *ap* conformer, and therefore, the addition of $Ba^{2+}$ shifts the equilibrium in favor of the *sp* conformer, leading to changes in the average chemical shifts of $L_1$ and $L_2$. These variations are predictable by analysis of the position of $L_1/L_2$ with respect to the phenyl ring of the MPA part in the *sp* conformation (Figure 1.11).

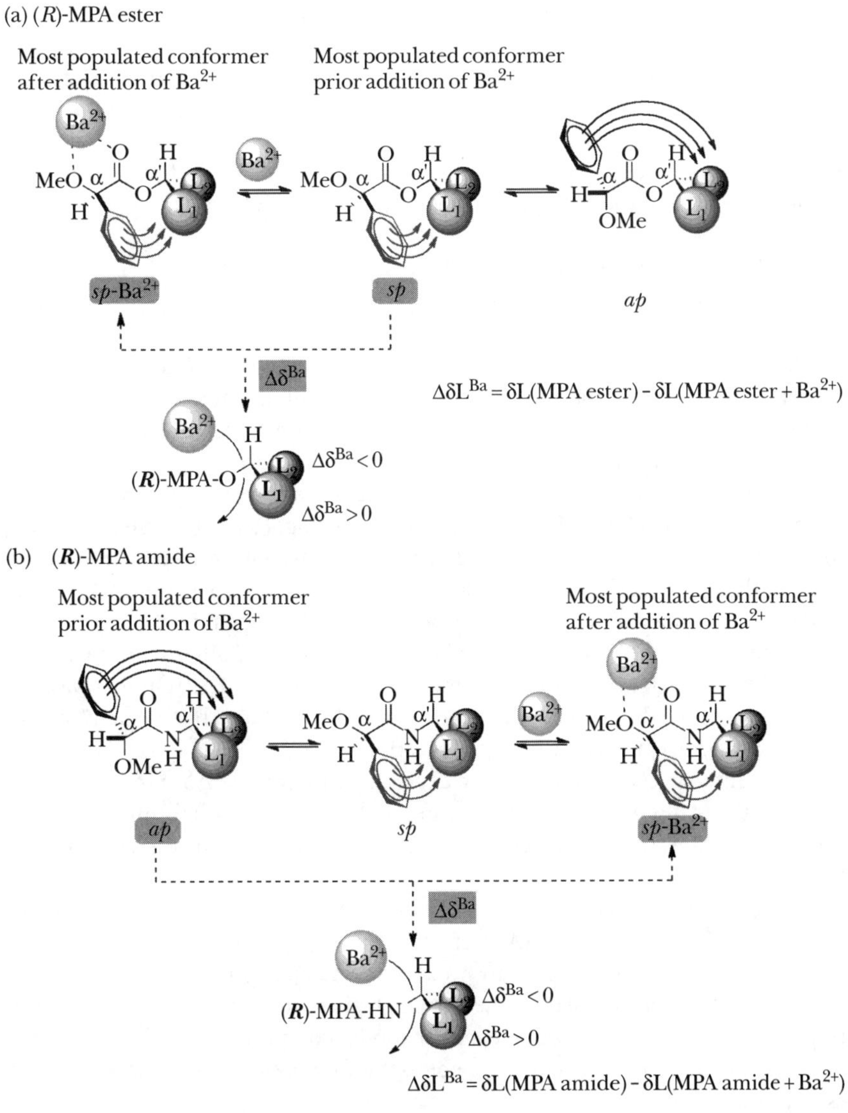

(a) (*R*)-MPA ester

$\Delta\delta L^{Ba} = \delta L(\text{MPA ester}) - \delta L(\text{MPA ester} + Ba^{2+})$

(b) (*R*)-MPA amide

$\Delta\delta L^{Ba} = \delta L(\text{MPA amide}) - \delta L(\text{MPA amide} + Ba^{2+})$

*Figure 1.11.* Conceptual representation of the single derivatization procedure, based on complexation with a metal cation.

Thus, there is a correlation between the greater or lesser shielding observed for $L_1/L_2$ after addition of $Ba^{2+}$ and their spatial location; we express this for each L substituent as $\Delta\delta^{Ba}$ = δ in the absence of $Ba^{2+}$ minus δ in the presence of $Ba^{2+}$. The sign of this difference, positive for one L and negative for the other, informs us about the location of the L with respect to the phenyl ring of the MPA in the *sp* conformation. Thus, it can be used for assigning the absolute configuration.

In this way, the assignment of alcohols or amines can be carried out using only one MPA derivative [42, 55–56]. The advantages of these single-derivatization procedures are that we use only half of the usual amount of substrate and have to carry out only one derivatization instead of two. However, there are also some limitations that should be considered.

For instance, in the case of the low-temperature method [41], one should check the solubility of the CDA derivative at low temperatures and avoid precipitation. In the case of adding $Ba^{2+}$, the major limitation is the complexation of $Ba^{2+}$ with the donor atoms located in the $L_1/L_2$ chains, in addition to or instead of those found in MPA. If that were to happen, one cannot be sure about the conformation of the complex, and therefore, no reliable assignment can be made.

### 1.6.3. Single-Derivatization Method Based on Esterification Shifts

In the particular case of secondary alcohols, there is another [43] single-derivatization method that requires no special solvent, metal ion, or temperature changes, but instead uses 9-AMA as the CDA (Figure 1.12). This method compares the chemical shifts observed for $L_1/L_2$ in the free alcohol and in the (*R*)- or (*S*)-9-AMA ester derivative [43]. The procedure is based on the known conformational composition of 9-AMA ester derivatives (analogous to MPA esters; *sp* being the main conformer) and the very strong shielding effect generated by the anthryl ring. In this way, the L substituent located on the same side as the anthryl ring in the *sp* conformer is very strongly shielded from the signal in the free alcohol ($L_1$ in the example of Figure 1.12), while the protons of the other L group are less shielded from the free alcohol ($L_2$ in the example of Figure 1.12). In this way, the magnitudes of the differences of the chemical shifts produced by the esterification with 9-AMA correlate with the spatial positions of $L_1/L_2$.

These differences are conveniently expressed through the $\Delta\delta^{AR}$ [δ in the alcohol minus δ in the (*R*)-9-AMA ester] or $\Delta\delta^{AS}$ [δ in the alcohol minus δ in the (*S*)-9-AMA ester] values, depending on which enantiomer of the auxiliary we have used [43].

Figure 1.12 shows schematically these differences in the spectra for the case of a free alcohol and its (*R*)-9-AMA ester, as well as a graphical model indicating the correlation between the signs of $\Delta\delta^{AR}$ and the spatial positions of $L_1/L_2$.

As the procedure is based on the very high shielding produced by the anthryl ring, other CDAs with the same conformational composition as MPA but without such a strong shielding effect are not useful for making assignments based on esterification shifts [43].

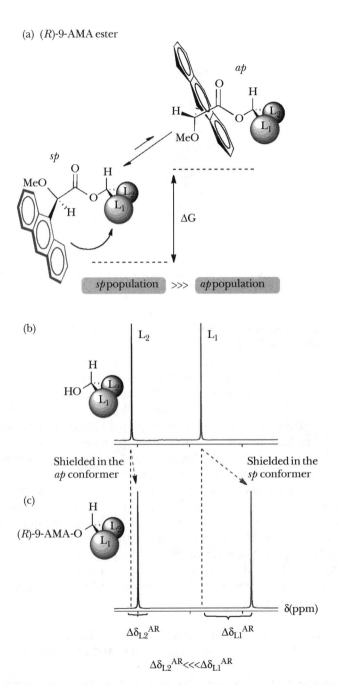

*Figure 1.12.* Conceptual representation of the single-derivatization procedure, based on esterification shifts.

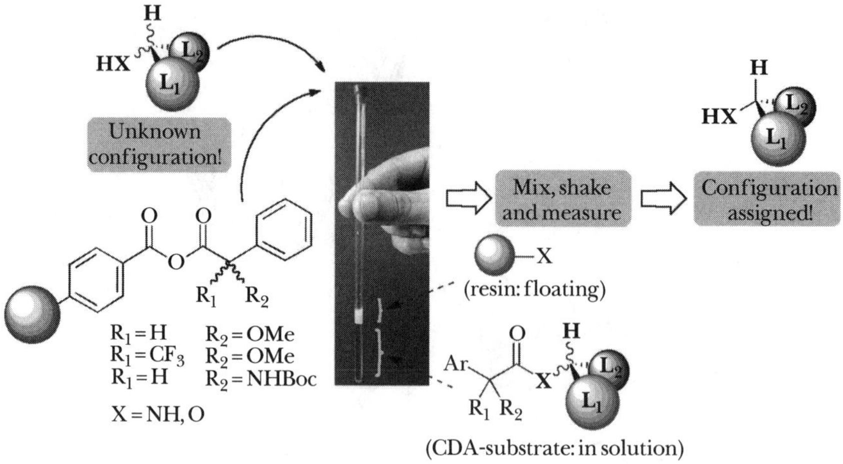

*Figure 1.13.* Representation of the mix-and-shake procedure, based on the use of resin-bound CDAs.

### 1.6.4. Mix-and-Shake Method: Assignment Using Resin-Bound CDAs

Assignments can be simplified by using only one derivative for only six classes of compounds: secondary alcohols [41–43], α-chiral primary amines [55–56], and *sec/sec*-1,2-diols [61], *prim/sec*-1,2-diols [65], and *prim/sec*- and *sec/prim*-1,2-amino alcohols [68]. Another way to simplify the assignment procedure is to incorporate the CDA reagent into a solid resin [83, 84]—a polystyrene similar to Merrifield resin—through a bond that allows the CDA moiety to act as an electrophile and the resin to act as a leaving group. In these conditions, the mixing in the NMR tube of the substrate (nucleophile) with the resin-bound CDA (plus resin scavengers, in some cases) produces the required CDA derivative, which remains in solution and ready to be analyzed by NMR; the resin-leaving group stays in the tube as a floating solid that does not interfere with the spectrum. See Figure 1.13.

This is the mix-and-shake procedure, which requires only mixing and shaking of the reagents in the NMR tube; there is no benchwork necessary for the preparation or purification of the derivatives. It is completely general for all classes of compounds listed in Figure 1.4.

## ■ 1.7. GENERAL CRITERIA FOR THE CORRECT APPLICATION OF THE NMR METHODOLOGY

As we have shown, the NMR methodology for configurational assignment requires [13–15]:

1) The derivatization of the substrate with the (*R*)- and the (*S*)-enantiomer of an adequate CDA (double derivatization) or with just one (single derivatization).

2) To record the $^1$H and/or $^{13}$C-NMR spectra, the assignment of the signals of $L_1$ and $L_2$ and the comparison of their chemical shifts in the two derivatives leading to $\Delta\delta^{RS}$ (double derivatization) or in two experimental conditions (two temperatures; without/with esterification; without/with $Ba^{2+}$) selected for the single-derivatization method leading to $\Delta\delta^{T1T2}$, $\Delta\delta^{AR}$, $\Delta\delta^{AS}$, or $\Delta\delta^{Ba}$.

3) The comparison of the $\Delta\delta$ sign distribution obtained experimentally with the graphical scheme (stereomodel) that illustrates the correlation between the spatial location of the $L_1/L_2$ substituents and their signs.

A couple of rules guarantee the correctness of the assignment [82]:

1. In order to make the $\Delta\delta^{RS}$ sign of each proton fully significant, the corresponding value should be higher than the experimental error of the chemical shift measurement; the identification of the signals due to $L_1/L_2$ should be correct, with no ambiguities; and the spectra should be recorded in the established experimental conditions (solvent, temperature).
2. All the proton and carbon nuclei in an L substituent (e.g., $L_1$) should have the same $\Delta\delta$ sign, and all the nuclei in the other L substituent (e.g., $L_2$) should have the opposite sign.

The reasons for the first point are easily deduced: a change in solvent (polar/donor effects) or temperature may modify the conformational equilibrium and therefore disturb the conformational distribution (i.e., *sp/ap* in MPA ester derivatives) that lies at the heart of this methodology. If some of the $\Delta\delta^{RS}$ values were too small to provide significant signs, the auxiliary reagent can be exchanged for another that provides a more intense shielding effect [37, 39] and, consequently, valid values and signs.

As for the condition indicated in the second point, some additional comments may be necessary, because in certain cases, the experimental signs either do not fully comply with those conditions or are incomplete, which may lead to erroneous assignments [82].

The cases that deserve special attention—because the reliability of the assignment derived from $\Delta\delta^{RS}$ signs is questionable—are summarized here [82]:

a) Identical signs of $\Delta\delta^{RS}$ are obtained for the $L_1/L_2$ substituents (Figure 1.14a).
b) Both positive and negative $\Delta\delta^{RS}$ signs are found within a substituent (Figure 1.14b).
c) Only one substituent provides $\Delta\delta^{RS}$ signs (Figure 1.14c).
d) Substrates contain more than one auxiliary reagent (Figure 1.14d).

The first case (a)—identical signs for $L_1/L_2$—occurs rarely, as might be expected, but it makes the assignment impossible. The only solution is to replace the CDA with another one and hope that a correct sign distribution will be obtained.

Case (b) represents a more common situation, where one of the L substituents presents a conformational mobility or a stereochemistry in which some nuclei are outside the shielding cone that affects the rest of the substituents. The importance of this lack of homogeneity in the signs depends on the number of nuclei involved and their distance to the asymmetric center. For instance, if only one proton at the

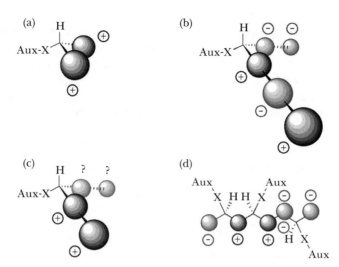

*Figure 1.14.* Possible scenarios where the assignments derived from the signs of $\Delta\delta^{RS}$ are questionable.

end of a long chain presents an unexpected sign, but the rest of the chain presents a homogeneous sign distribution, then one could guess that the "abnormal" sign is not very important and carry out the assignment while disregarding that isolated point. Naturally, one way to increase confidence in the assignment would be, as before, to exchange the CDA for another one that is more intense and/or more adequate for long-distance nuclei and then recheck the signs.

Case (c) was very important when the assignment was based only on $^1$H NMR, because one can find a number of chiral compounds with no protons in one of the $L_1/L_2$ substituents. Therefore, only the $\Delta\delta^{RS}$ signs in one substituent are obtained, and no sure assignment is possible. A good example is an aldehyde cyanohydrin in which only one of the L substituents gives proton signals, because the other L substituent is the CN group. In general, this limitation disappears when using $^{13}$C NMR for the assignment. This means that in those cases, we should determine the $^1$H and $^{13}$C-NMR spectra, and then use the $\Delta\delta^{RS}$ signs of both protons (the L chain in the cyanohydrine) and the carbons (the CN in the cyanohydrin) as if they originated from a single NMR spectrum.

Finally, case (d) occurs very frequently, and, in particular, it always happens when the substrate has more than one derivatizable group [13, 82]. Imagine your substrate is a *sec/sec*-1,3-diol with two asymmetric carbons [59, 60]. Naturally, it is very difficult to prepare the mono AMAA derivatives, because both OH groups will react with the AMAA reagent at basically the same rate; it is better to functionalize the two OH groups at the same time with the (*R*)- and the (*S*)-AMAA, thus producing the corresponding bis-(*R*)- and bis-(*S*)-AMAA derivatives.

As the structure of the diol resembles a dimer of the mono alcohol, where a chain links two monoalcohol units, researchers tried for years to assign the absolute configuration of the diol as if the two alcohol moieties were independent.

This is not correct, because the chemical shifts respond to the joint action of the aromatic shieldings produced by the two AMAA auxiliaries [59, 60] (Figure 1.15), and therefore, they do not correlate with the shielding produced by just one AMAA unit. In fact, for compounds with more than one derivatizable group, the correlation between the stereochemistry and the $\Delta\delta^{RS}$ signs responds to specific models [13, 59–70] in ways that are different from those of the mono AMAA derivatives. This will be shown in the next section.

## ■ 1.8. CORRELATION MODELS FOR THE ASSIGNMENT OF POLYFUNCTIONAL COMPOUNDS

As mentioned above, when there is more than one auxiliary group (CDA) in the molecule (e.g., two), and those auxiliaries are not very far one from the other, the chemical shifts observed for the bis-CDA derivatives reflect the combination of the shieldings caused by both CDA units [13]. This combination results in unpredictable $\Delta\delta^{RS}$ signs for the $R_1/R_2$ substituents (Figure 1.15) [13, 59, 60]. Therefore, those protons are useless for making assignments. For instance, if one proton is shielded by one CDA unit in the bis-(R)-derivative and shielded by the other CDA unit in the bis-(S)-derivative (Figure 1.15), the resulting $\Delta\delta^{RS}$ will be the difference between two close values and thus too small to produce a significant sign.

The complexity of these combinations of aromatic shielding effects and their variations with different stereochemistry is beyond the scope of this book (a more detailed explanation can be found in Chapter 5). It should be enough to say that the deduction of the correlation between NMR spectra and stereochemistry [13] follows the same methodology as that of the monoderivatized compounds [14]: study of the main conformations, orientation, and intensity of the aromatic shielding effects; identification of the protons whose shifts respond to the stereochemistry; and testing with a collection of compounds of known absolute configuration. In this way, the correlation between the signs of the NMR shifts and the absolute stereochemistry can be deduced for those polyfunctional compounds and then used for assignment purposes [13, 59–70].

It should be mentioned that the correlation models for polyfunctional compounds are much more complex than those for monofunctional ones, and in general, they cannot be represented by just a tetrahedron. This is because: a) the signals useful for assignment are not necessarily those of $R_1/R_2$; and b) if more than one

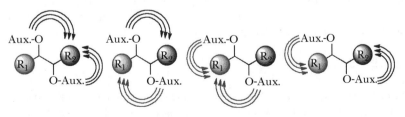

*Figure 1.15.* Possible combinations of aromatic shieldings produced by two AMAA auxiliaries in a 1,2-diol.

## $\Delta\delta^{RS}(\Delta\delta^{SR})$ Sign distribution[a]

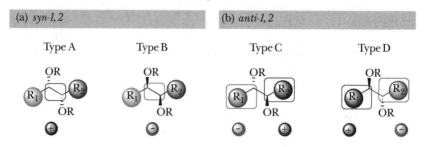

[a]Sign distribution: $\Delta\delta^{RS}$, CDA = MPA, 9-AMA, 1-NMA, 2-NMA; $\Delta\delta^{SR}$, CDA = MTPA

*Figure 1.16.* Correlation between the absolute stereochemistry of *sec/sec*-1,2-diols derivatized as bis-AMAA esters and the $\Delta\delta^{RS}$ signs ($\Delta\delta^{SR}$ for bis-MTPA derivatives).

asymmetric carbon is present in the substrate, more than two isomers are possible (e.g., in a *sec/sec*-diol, there are four stereochemical possibilities), and the NMR method should be able to produce four different sign distributions (Figure 1.16).

As an example, Figure 1.16 shows the correlation between the absolute stereochemistry of *sec/sec*-1,2-diols derivatized as bis-AMAA esters and the $\Delta\delta^{RS}$ signs ($\Delta\delta^{SR}$ for bis-MTPA derivatives) of H$\alpha$(R$_1$), H$\alpha$(R$_2$), R$_1$, and R$_2$. As can be seen, there are four different combinations of signs, each one specific to a possible stereoisomer. The configurations of the two asymmetric carbons are assigned in the same operation (Figure 1.16).

Other differences from monofunctional compounds [14] are that only four classes of compounds—*sec/sec*-1,2-diols [61], *prim/sec*-1,2-diols [65], *prim/sec*-1,2-amino alcohols and *sec/prim*-1,2-amino alcohols [68]—can be assigned by low-temperature NMR and that the very small $\Delta\delta$ $^{13}$C-NMR shifts obtained due to the aforementioned combination of shieldings renders useless the use of $^{13}$C NMR with polyfunctional substrates [72].

Nevertheless, there is a fairly wide variety of polyfunctional compounds whose absolute configurations can be deduced using the NMR methodology presented in this book [13] (*sec/sec*-1,2-diols [59,61], *sec/sec*-1,2-amino alcohols [62], *prim/sec*-1,2-diols [63–65], *prim/sec*-1,2-amino alcohols [66, 68], *sec/prim*-1,2-amino alcohols [66, 68], and *prim/sec/sec*-1,2,3-triols [69, 70], Figure 1.4). Figure 1.17 shows their respective structures, their diagnostic signals, and the adequate CDAs for each class. The actual correlation models used for assignment are included with the applications presented in Chapter 5.

## ■ 1.9. SUMMARY

The conclusion we can draw from the previous sections is that, for certain organic compounds, it is possible to carry out the assignment of their absolute configuration by examination of the NMR spectra of certain derivatives prepared when using chiral AMAAs as CDAs [13–15].

The Theoretical Basis for Assignment by NMR ■ 23

| Substrate | | Diagnostic protons | Derivatization method (CDA) |
|---|---|---|---|
| syn-1,2-diols | 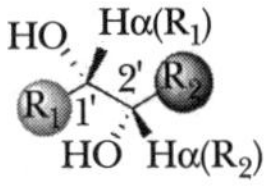 HO Hα(R₁) / R₁–1'–2'–R₂ / HO Hα(R₂) | Hα(R₁), Hα(R₂) | Double (MPA, 9-AMA, MTPA) |
| anti-1,2-diols | 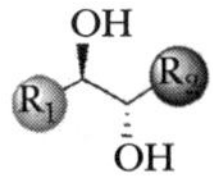 OH / R₁–R₂ / OH | Hα(R₁), Hα(R₂), R₁, R₂ | Double (MPA, 9-AMA, MTPA) |
| | | R₁, R₂ | Single (MPA) |
| syn-1,2-amino alcohols | 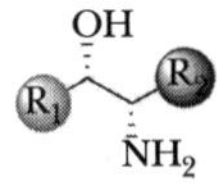 OH / R₁–R₂ / NH₂ | Hα(R₁), Hα(R₂), R₁, R₂ | Double (MPA) |
| anti-1,2-amino alcohols | 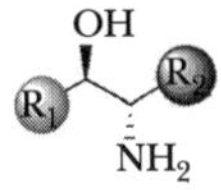 OH / R₁–R₂ / NH₂ | Hα(R₁), Hα(R₂) | Double (MPA) |
| prim/sec-1,2-diols | 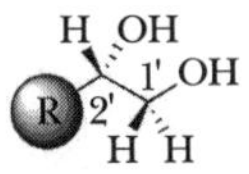 H OH / R–2'–1'–OH / H H | CH₂(1'), R | Double (MPA, 9-AMA) |
| | | CH₂(1') | Single (MPA, 9-AMA) |
| sec/prim-1,2-amino alcohols | 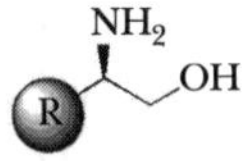 NH₂ / R–OH | CH₂(1'), R | Double (MPA) |
| | | (CαH, OMe)$_{MPA}$ | Double (MPA) |
| | | (CαH, OMe)$_{MPA}$ | Single (MPA) |
| prim/sec-1,2-amino alcohols | 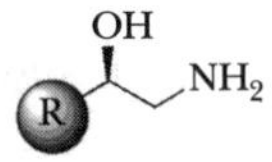 OH / R–NH₂ | CH₂(1'), R | Double (MPA) |
| | | (CαH, OMe)$_{MPA}$ | Double (MPA) |
| | | (CαH, OMe)$_{MPA}$ | Single (MPA) |
| prim/sec/sec-1,2,3-triols | 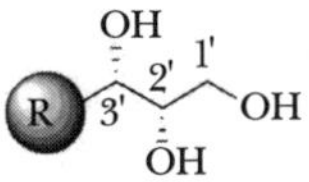 OH 1' / R–3'–2'–OH / OH | H(2'), H(3') | Double (MPA) |

*Figure 1.17.* Recommended CDAs for the assignment of polyfunctional substrates, together with the protons whose signals are used for diagnosis and the derivatization method employed.

The method is based on the aromatic shielding effect that is produced by the CDA part of the derivatives on the chemical shifts of the substrate; evidently, these are very dependent on the conformational composition of the derivatives. Therefore, different classes of substrates may require different CDAs to provide the most reliable data and assignments.

Figures 1.17 and 1.18 present the CDAs that are best suited for making assignments for several classes of poly- [14] and monofunctional [13] substrates; also included are the protons whose signals are used for the diagnosis and the derivatization method employed.

It should be stressed at this point that the correlation between the spatial locations of $L_1/L_2$ and the NMR shifts indicated by these graphical models, have been not only deduced from experimental and theoretical data (e.g., conformational analysis and theoretical calculations, including geometry minimizations and aromatic shielding effects), but also validated with a number of compounds of known

24 ■ The Assignment of the Absolute Configuration by NMR

| Substrate | | Diagnostic protons | Derivatization method (CDA) |
|---|---|---|---|
| Secondary alcohols | 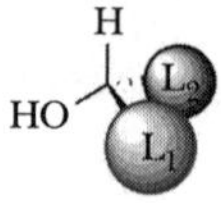 HO–C(H)(L₂)(L₁) | L₁, L₂ | Double (MPA, 9-AMA, MTPA, AMAAs) Single (MPA, 9-AMA) |
| Primary alcohols | 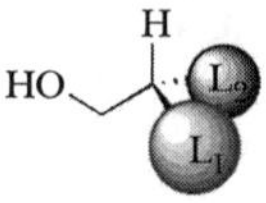 HO–C(H)(L₂)(L₁) | L₁, L₂ | Double (9-AMA) |
| Ketone cyanohidrins | 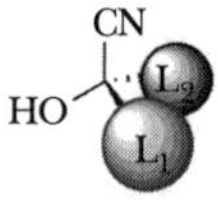 HO–C(CN)(L₂)(L₁) | L₁, L₂ | Double (MPA) Single (MPA) |
| Aldehyde cyanohidrins | 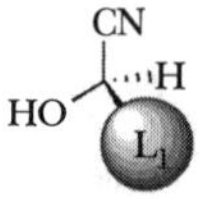 HO–C(CN)(H)(L₁) | L₁, CN | Double (MPA) |
| Primary amines | 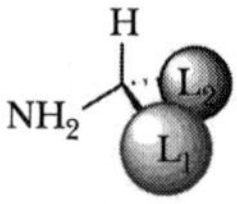 NH₂–C(H)(L₂)(L₁) | L₁, L₂ | Double (BPG, MPA, MTPA) Single (MPA) |
| Secondary thiols | 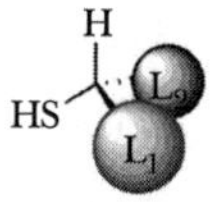 HS–C(H)(L₂)(L₁) | L₁, L₂ | Double (MPA, 2-NTBA) |
| Carboxylic acids | 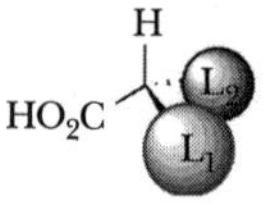 HO₂C–C(H)(L₂)(L₁) | L₁, L₂ | Double (9-AHA) |

*Figure 1.18.* Recommended CDAs for the assignment of monofunctional substrates, together with the protons whose signals are used for diagnosis and the derivatization method employed.

absolute configuration that were consistent with predictions in all cases. The sensitivity of the conformational composition to changes in the substrate or the experimental NMR conditions makes this testing essential for the reliability of the procedure. The ample information available on arylalcoxyacetic acids and their derivatives is the main reason for their selection for NMR assignment; many other CDAs are described in the literature but have not been sufficiently tested with compounds of known absolute configuration to demonstrate the general character and validity of the assignments derived from them.

The same warning can be made when a CDA/procedure, well proven for the assignment of a particular class of compounds, is automatically applied to a different class of substrate without any experimental or theoretical study that might prove its validity.

In Chapter 2, we will show the experimental details related to the preparation/source of the CDAs, of the CDA derivatives, the recovery of the substrate (when possible), the best conditions for the NMR experiments, and the simplified procedures (such as the use of resin-supported CDAs) that have been devised to reduce the necessary benchwork.

In Chapters 3–5, we will describe the methods and reagents for a series of different substrates and show examples and spectra for applications of these procedures.

The last chapter (Chapter 6) contains a collection of 50 problems selected to train the reader to use this methodology.

# 2 Practical Aspects of the Preparation of the Derivatives

## ■ 2.1. INSTRUMENTATION, CONCENTRATION, SOLVENT, AND TEMPERATURE OF THE NMR EXPERIMENT

Most of the NMR spectra shown in this book and in the literature have been recorded at 250 or 300 MHz, with a few being obtained at 500 MHz for $^1$H NMR (the equivalent for $^{13}$C NMR). No special pulse sequences are necessary, just standard one-dimensional (1D) spectra although two-dimensional (2D) experiments (e.g., correlation spectroscopy; COSY) may be necessary in some cases in order to get an unambiguous identification of the signals relevant for the assignment.

In general, 5–10 mg or less of CDA derivative dissolved in 0.5 mL of deuterated solvent are sufficient to obtain a good NMR spectrum.

Temperature, solvent, and concentration used in the NMR experiments should be adequate for each CDA-substrate pair and methodology, because the method is based on the conformational composition of the AMAA derivatives in precise conditions.

With the exception of the low-temperature procedure (single derivatization), a NMR probe temperature around 300 K has always been used.

In general, the spectra for double-derivatization assignments should be taken in deuterated chloroform. Different NMR solvents are required only in two of the single-derivatization methods.

In the assignment by low-temperature NMR, the most convenient solvent is a $CS_2/CD_2Cl_2$ (4:1) mixture, which allows the use of temperatures low enough (i.e., 213 K) to obtain relevant shifts. In the procedure based on the complexation with $Ba^{2+}$, the NMR solvent should be deuterated acetonitrile. The barium salt is anhydrous $Ba(ClO_4)_2$, which can be added directly to the tube by using a spatula. No weighing is necessary after shaking, as the excess salt will remain at the bottom of the NMR tube and will not disturb the experiment.

## ■ 2.2. SOURCE AND PREPARATION OF THE CDAS

(*R*)- and (*S*)-MPA, MTPA, and Boc-phenylglycine (BPG) are commercially available and can be used without further purification. The first two (MPA and MTPA) can also be purchased as acid chlorides. When using MTPA or the corresponding acid chloride [85] for the derivatization of an alcohol or amine, it should be noted that the Cahn-Ingold-Prelog priority rules assign different *R/S* descriptors to the acid and to the corresponding chloride; this is due to the different priority order generated by the substituents [i.e., (R)-MTPA generates the (S)-acid chloride and vice versa].

The other auxiliary reagents described in this book (2-naphthylmethoxyacetic acid, 2-NMA; 1-NMA; 9-AMA; 2-(*tert*-butoxy)-2-(2-naphthyl)acetic acid, 2-NTBA; and 9-AHA) are not commercially available, but they can be easily prepared in chiral form, as described in the literature [37, 50, 57, 64, 86–90].

## 2.3. PREPARATION OF THE CDA ESTERS, THIOESTERS, AND AMIDES

The preparation of the CDA derivatives is based on standard reactions for esters, amides, and thioesters, from free acids or acid chlorides. The experimental details mentioned here have been optimized to reduce to a minimum the number of purification/isolation steps. When a derivatization was carried out with a carboxylic acid as the CDA, we used either excess acid (1.2 equiv) and/or the addition of a coupling agent (e.g., *N,N*'-dicyclohexylcarbodiimide, DCC; 4-dimethylaminopyridine, DMAP) in order to accelerate the reaction and the obtention of better yields. The derivatizations of alcohols and amines with CDA acid chlorides are fast enough when carried out in pyridine and thus do not require coupling reagents.

Due to the excess CDA reagent and the use of DMAP, *N,N*-diisopropylethylamine (DIPEA), DCC, pyridine, or other agents, some purification of the derivatives is necessary before a good NMR spectrum can be obtained. In our experience, those substances can be eliminated by using flash chromatography on a short silica gel column (i.e., on a Pasteur pipette) and eluting with hexane-ethyl acetate mixtures.

Standard experimental procedures taken from the literature follow (Sections 2.3.1 through 2.3.4). There are other procedures, analogous to those discussed in Sections 2.3.1 through 2.3.4, for di- and trifunctional substrates. These merely require adjusting the amount of CDA and coupling agents.

### 2.3.1. Derivatization of Alcohols, Thiols, and Cyanohydrins Using the CDA Acid

A solution of the substrate (alcohol, thiol, cyanohydrin; 0.5 mmol) in 10 mL of $CH_2Cl_2$ is treated with the auxiliary acid (MPA, 9-AMA, or other agent; 0.6 mmol) in the presence of DCC (0.6 mmol) and DMAP (cat). When the reaction is complete (monitoring with thin-layer chromatography, TLC), the mixture is filtered to remove the dicyclohexylurea, and the CDA-ester is then purified by flash chromatography on silica gel and eluted with hexane-ethyl acetate mixtures [15, 91].

### 2.3.2. Derivatization of Amines Using the CDA Acid

To a solution of the free amine or the amine salt (0.5 mmol) dissolved in 10 mL of $CH_2Cl_2$, the CDA acid (MPA, MTPA, BPG, or other agent; 0.6 mmol) and 0.6 mmol of DCC (free amine) or 0.6 mmol of DCC and 0.5 mmol of DMAP (amine salt) are added. The reaction mixture is filtered to remove the dicyclohexylurea, and the CDA-amide is then purified by flash chromatography on silica gel and eluted with dichloromethane [15, 52, 91].

### 2.3.3. Preparation of the CDA Acid Chlorides

To a mixture of the chiral CDA acid (MPA, 9-AMA, or other agent; 0.5 mmol) and dimethylformamide (DMF; 0.05 mmol) in hexane (5 mL), oxalyl chloride (5.0 mmol) is added at room temperature. After 2 days, the solvent is evaporated to dryness and the residual acid chloride is used without further purification [15, 92, 93].

### 2.3.4. Derivatization of Alcohols, Thiols, and Cyanohydrins Using the CDA Acid Chloride

A solution of the CDA acid chloride (MPA-Cl, 9-AMA-Cl, or other agent; 0.5 mmol; prepared as described in Section 2.3.3 and dissolved in 5 mL of $CH_2Cl_2$) is added to a solution of the substrate (alcohol, thiol, cyanohydrin; 0.1 mmol) in 10 mL of dry $CH_2Cl_2$ containing $Et_3N$ (1.2 mmol) and DMAP (0.1 mmol). After stirring for 15 min, the solution is washed with $H_2O$, dried, and concentrated to a residue that is purified by flash chromatography on silica gel [15, 92, 93].

## ■ 2.4. RESIN-BOUND CDA DERIVATIVES (MIX-AND-SHAKE METHOD)

The mix-and-shake method for the assignment of absolute configuration [83, 84] involves the use of auxiliary chiral agents (CDAs) anchored to polymer resins. Its use results in significant progress in the preparation of the derivatives, because no benchwork is necessary: just mix the substrate, the solvent, and the CDA-linked resin in the NMR tube, and shake it for several minutes (Figure 2.1).

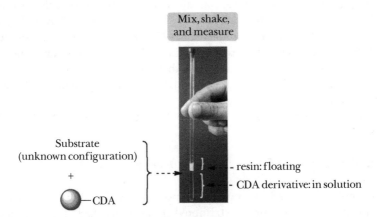

*Figure 2.1.* Mix-and-shake method for the assignment of absolute configuration.

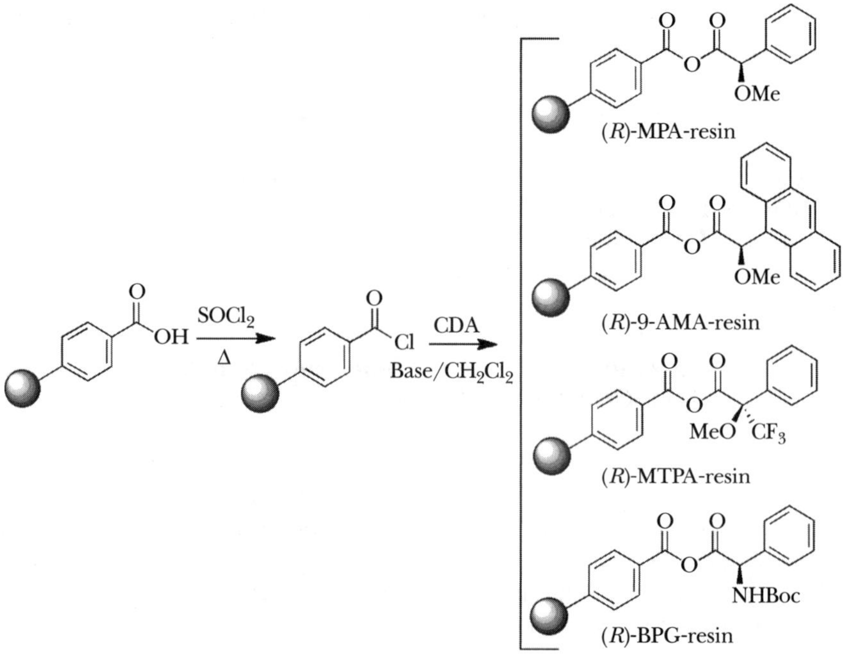

*Figure 2.2.* Polystyrene resins bound to common CDAs through mixed anhydride bonds.

### 2.4.1. Preparation of Resin-Bound CDA Derivatives

Several resins can be used to link the most common CDAs, but in this book we will focus only on polystyrene-type resins bound to the CDAs through a mixed anhydride bond (Figure 2.2).

The starting polymer is a commercial carboxypolystyrene hybrid-layer (HL) resin [100–200 mesh; loading 1.4 mmol/g and with 1% divinylbenzene (DVB) crosslinking]. The resin-bound CDAs are prepared by transforming the carboxylic group of the starting polymer into an acid chloride (using oxalyl or thionyl chloride); this then reacts with MPA, MTPA, BPG, or another CDA. In this way, the resin-bound (R) and (S) enantiomers of a CDA can be prepared and stored for months before use. In some cases, it is useful to obtain resins containing both (R) and (S) enantiomers of the CDA in a ratio that is known and is different from 1:1 [e.g., (R)-MPA and (S)-MPA in a 3:1 ratio]. These can be prepared by using the required mixture of the CDA enantiomers.

Standard procedures taken from the literature follow in Sections 2.4.2 through 2.4.9.

### 2.4.2. Preparation of Acid Chloride Resins

Carboxypolystyrene resin (500 mg, 2 mmol) in dry $CH_2Cl_2$ (2 mL) is slowly stirred under Ar for 30 min. Next, thionyl chloride (6 mL, 82 mmol) is added, and the resulting mixture is heated at 65 °C for 4 h. After this, the resin is filtered under Ar, washed with dry $CH_2Cl_2$ (8 × 3 mL), and dried overnight under vacuum [83, 84].

### 2.4.3. Preparation of CDA-Resins

Acid chloride resin (500 mg, 1.9 mmol) in dry $CH_2Cl_2$ (2 mL) is slowly stirred under a flow of Ar (30 min). The required enantiomer of the CDA (MPA, 9-AMA, BPG or MTPA; 2.3 mmol) and dry DIPEA (2.3 mmol) dissolved in dry $CH_2Cl_2$ (2 mL with MPA, BPG, and MTPA; 4 mL with 9-AMA) are then added to the resin. The resulting mixture is stirred under Ar (1 h with MPA, BPG, and MTPA; 6 h with 9-AMA), and then the resin is filtered off, washed with dry $CH_2Cl_2$ (8 × 3 mL), dried overnight under vacuum, and stored at −22 °C under an Ar atmosphere (the formation of the anhydride can be monitored by on-bead infrared microscopy, IR). At room temperature, these resins are also stable when kept dry under vacuum, showing less than 5% cleavage of MPA after 4 months storage [83, 84].

### 2.4.4. Determination of the Loading of the CDA-Resins

The CDA resin (30.0 mg), toluene (75 μmol, 6.9 mg), and dry benzylamine (225 μmol, 24.1 mg) are introduced into an NMR tube containing $CDCl_3$ (250 μL). The mixture is shaken (2 h), extra $CDCl_3$ (350 μL) is added, and the $^1$H-NMR spectrum is recorded. The average loading is calculated by integration of the amide signals with respect to those of toluene, which serves as an internal reference. Resins prepared according to this protocol showed the following loadings: 2.2 mmol/g for the MPA-resin; 1.7 mmol/g for the MTPA- and BPG-resins, and 1.4 mmol/g for the 9-AMA-resin [83, 84]. All of the experiments described in Sections 2.4.5 through 2.4.9 have been carried out using the CDA-resins and loading described here.

### 2.4.5. In-Tube Derivatization of Amines

The reactivity of primary amines as nucleophiles is fast enough that the reaction with resin-bound MPA, or BPG is completed within a few minutes. No addition of coupling reagents is necessary.

The derivatization is carried out directly in the NMR tube by mixing together 300 μL of $CDCl_3$, the amine, and two to three times more resin-bound CDA than is necessary. A typical experiment follows.

CDA-MPA resin (44 μmol) and the amine (22 μmol) are added to the NMR tube containing dry $CDCl_3$ (150 μL). After shaking the mixture for 5–10 min, extra $CDCl_3$ (450 μL) is added, and the NMR spectrum is recorded. In the $CDCl_3$, the resin remains floating in the tube, but it does not interfere with the spectrum (Figure 2.3) [83, 84].

The derivatization of amines with resin-bound-MPA can also be carried out in $CD_3CN$, thus allowing a combination of the mix-and-shake and barium complexation procedures (Figure 2.4). In $CD_3CN$, the reaction is slower, and the mixture should be shaken for 30 min. After that time, extra $CD_3CN$ is added, and the first

32 ■ The Assignment of the Absolute Configuration by NMR

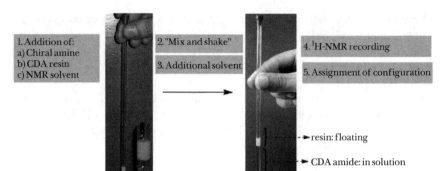

Figure 2.3. Mix-and-shake method for the assignment of the configuration of amines.

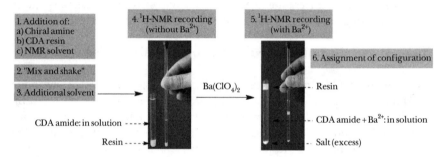

Figure 2.4. Combination of the mix-and-shake and the barium complexation methods for the assignment of the configuration of amines.

spectrum is recorded. In $CD_3CN$, the resin falls to the bottom of the tube instead of floating, but again, it does not interfere with the experiment [83, 84]. Solid $Ba(ClO_4)_2$ is then added, and the tube is shaken for 5 min; this allows the second spectra to be obtained. Now, the resin has recovered its floating properties.

### 2.4.6. In-Tube Derivatization of Primary and Secondary Alcohols, Cyanohydrins, and Secondary Thiols

Because alcohols and thiols are weaker nucleophiles than amines, their derivatization is slower and can take several hours to be completed. This time can be reduced to just 5–10 min if coupling reagents (e.g., DMAP) are added to the NMR tube. Naturally, the resulting reaction mixture contains, in addition to the desired CDA derivative, DMAP and some free CDA produced by hydrolysis of the initial resin-bound CDA reagent. These byproducts can be easily removed from the solution by the addition to the NMR tube of two scavenger resins: a sulphonic acid resin [94] to trap the DMAP and an amino resin [95, 96] to trap the free CDA (Figure 2.5). After 1 h of shaking, the tube can be submitted to the NMR experiment, and it will show a clean spectrum corresponding to the CDA derivative [83, 84].

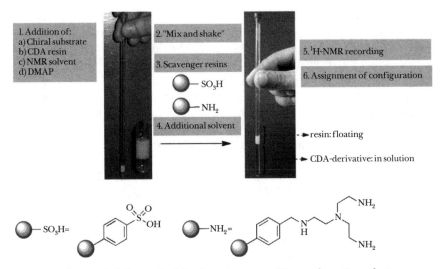

*Figure 2.5.* Mix-and-shake method for the assignment of the configuration of primary and secondary alcohols, cyanohydrins, secondary thiols, amino alcohols, diols, and triols.

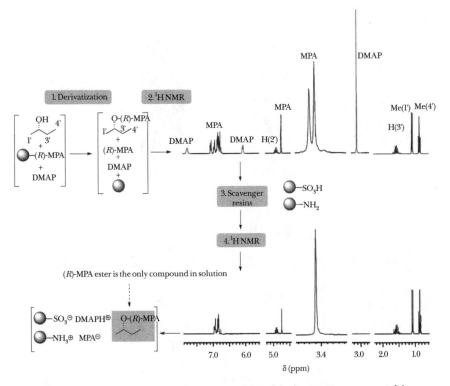

*Figure 2.6.* Mix-and-shake method for the recording of the ¹H-NMR spectrum of the (*R*)-MPA ester of (*S*)-butan-2-ol.

In a typical experiment, a solution of DMAP (22 µmol) in CDCl$_3$ (100 µL) is added to an NMR tube containing the CDA resin (44 µmol) and the substrate (22 µmol) in dry CDCl$_3$ (100 µL). After shaking for 10 min, amino and sulphonic acid scavenger resins (44 µmol of each) and extra CDCl$_3$ (200 µL) are added to the tube, and the shaking continues for 1 h. Following this, extra CDCl$_3$ (450 µL) is added, and the NMR spectrum is recorded [83, 84] (Figure 2.6).

### 2.4.7. In-Tube Derivatization of Amino Alcohols

MPA resin (66 µmol) and the amino alcohol (22 µmol) are added to an NMR tube containing dry CDCl$_3$ (200 µL). A solution of DMAP (22 µmol) in CDCl$_3$ (100 µL) is then added to the tube, and the mixture is shaken for 20 min. Following this, extra CDCl$_3$ (300 µL) is added, and then the NMR tube is ready for recording [83, 84] (Figure 2.6).

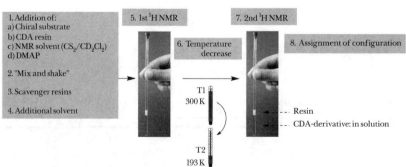

*Figure 2.7.* Mix-and-shake method for single-derivatization procedures (complexation with Ba$^{2+}$ and low temperature).

## 2.4.8. In-Tube Derivatization of Diols and Triols

A solution of DMAP (44 µmol) in $CDCl_3$ (100 µL) is added to an NMR tube containing the MPA-resin (66 µmol) and the diol (22 µmol) in dry $CDCl_3$ (200 µL). After the mixture is shaken for 90 min, extra $CDCl_3$ (300 µL) is added, and the NMR spectrum is recorded [83, 84] (Figure 2.6). Optionally, scavenger resins can also be used.

For triols, there is an analogous procedure, but the amount of the reagent and the reaction time are both different. Typical data are: 16.5 µmol of triol, 66 µmol MPA-resin, 33 µmol of DMAP, 8 h reaction time [83, 84].

## 2.4.9. In-Tube Derivatization for Single-Derivatization Procedures

The mix-and-shake in-tube preparation can also be used for the assignment of the absolute configuration based on the use of a single derivative for secondary alcohols, *sec/sec*-1,n- and *prim/sec*-1,2-diols, and *prim/sec*- and *sec/prim*-1,2-amino alcohols. The reaction between the substrate and the resin-bound CDA is carried out in the tube as usual, but particular solvents are used for each procedure. In the barium(II) complexation method for MPA derivatives of alcohols (and amines, see Section 2.4.5), the solvent of choice is $CD_3CN$ (Figure 2.7a). In the low-temperature approach, this is the $CS_2/CD_2Cl_2$ (4:1) mixture (Figure 2.7b). In both cases, the derivatization reaction is a little bit slower than in $CDCl_3$, requiring around 30 min for completion [83, 84].

# 3 Assignment of the Absolute Configuration of Monofunctional Compounds by Double Derivatization

## 3.1. SECONDARY ALCOHOLS

The assignment of secondary alcohols can be carried out by using one of several CDAs [13–15]. The most used and most reliable ones are MPA, 9-AMA, and MTPA [35–40]. Figure 3.1 shows their structures, the correlation models, and a summary of the experimental conditions. MPA and 9-AMA esters share the same conformational composition [37, 39] and only differ in the intensity of their shieldings; therefore both auxiliaries present the same correlation between sign distribution and stereochemistry. MTPA has a different conformational composition and correlation model [38].

### 3.1.1. MPA and 9-AMA as CDAs for Secondary Alcohols

As shown in Chapter 1, MPA esters of secondary alcohols and other AMAA esters (e.g., 9-AMA esters) are composed of two *sp/ap* conformers in fast equilibrium [37, 39]. The *sp* conformer is more stable than the *ap* conformer, and this allows the NMR spectrum of an AMAA ester to be interpreted as if it had originated from just the *sp* form: carbonyl, Cα, and methoxy groups in the auxiliary part and a methine group (Cα'-H) at the alcohol moiety are in the same plane (Figures 1.6 and 3.2).

When we consider this conformation in the (*R*)- and the (*S*)-AMAA esters, the $L_1$ group is located under the shielding cone of the aryl in the (*R*)-AMAA ester, while the $L_2$ is shielded in the (*S*)-AMAA ester (we strongly recommended that the reader builds Dreiding, or similar, models to assist in visualizing this spatial array).

A subtraction defined as the chemical shift in the (*R*)-AMAA ester minus that in the (*S*)-AMAA ester results in a negative value for $L_1$ and a positive one for $L_2$ (i.e., negative $\Delta\delta^{RS}$ for $L_1$ and positive $\Delta\delta^{RS}$ for $L_2$). Therefore, for any secondary alcohol derivatized as an AMAA ester, the protons showing a positive $\Delta\delta^{RS}$ sign are located in the tetrahedron around the asymmetric carbon (Cα') as $L_2$ (at the back) while the protons resulting in a negative $\Delta\delta^{RS}$ take the position of $L_1$ (at the front; Figure 3.2).

Thus, from a practical point of view [15], the assignment of configuration of a secondary alcohol requires the calculation of the $\Delta\delta^{RS}$ signs for the protons located

**38** ■ The Assignment of the Absolute Configuration by NMR

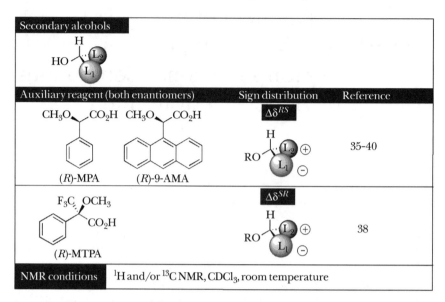

Figure 3.1. The assignment of chiral secondary alcohols at a glance: CDAs, sign distributions ($\Delta\delta^{RS}$, $\Delta\delta^{SR}$), experimental conditions, and references.

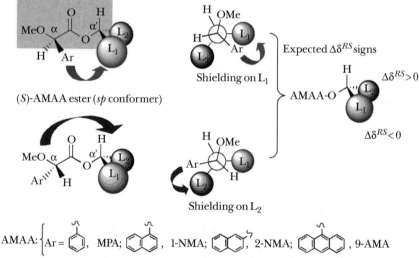

Figure 3.2. Shielding/deshielding effects in the most representative conformers of the AMAA esters, and the expected $\Delta\delta^{RS}$ signs.

at both sides of the plane defined in the *sp* conformation as well as their location at the front or the back of the tetrahedron, according to their $\Delta\delta^{RS}$ signs [39].

Of course, it is also possible to predict the distribution of $\Delta\delta^{RS}$ signs expected for a certain enantiomer of the alcohol just by building the Dreiding models representing the *sp* conformers [15, 37, 39]. The process for assignment is illustrated step by step in the following examples.

### 3.1.2. Example 1: Assignment of the Absolute Configuration of Diacetone D-Glucose Using MPA

A sample of a diacetone glucose of unknown configuration (L or D) is separately derivatized with (R)- and (S)-MPA, and the $^1$H-NMR spectra of the MPA esters are recorded (Figures 3.3a and b). The signals are assigned, paying special attention to the protons located at the two sides of the *sp* plane. In this case, the signals for diagnosis are those from the H(1')/H(2') on one side, and the H(4')/H(5')/H(6') at the other (Figure 3.3b).

The subtraction of the chemical shifts leads to the $\Delta\delta^{RS}$ signs shown in Figure 3.3c: negative for H(4')/H(5')/H(6') and positive for H(1')/H(2'). Thus, in accordance with the sign/spatial location correlation for secondary alcohols (Figure 3.3d), H(4')/H(5')/H(6') are located at the front ($L_1$ in Figure 3.3d) and H(1')/H(2') at the back of the tetrahedron around Cα' ($L_2$ in Figure 3.3d). These results lead to the stereochemistry shown in Figure 3.3d, which corresponds with that of diacetone-D-glucose; this is coincident with the known configuration of the initial sample.

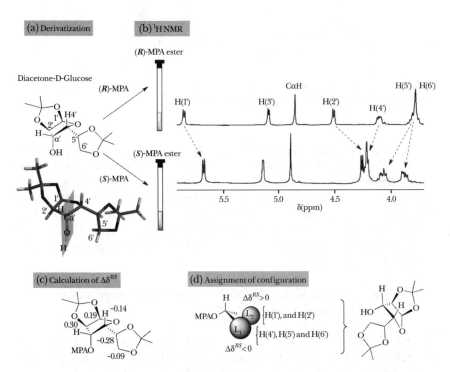

*Figure 3.3.* Main steps in the configurational assignment of diacetone D-glucose using MPA.

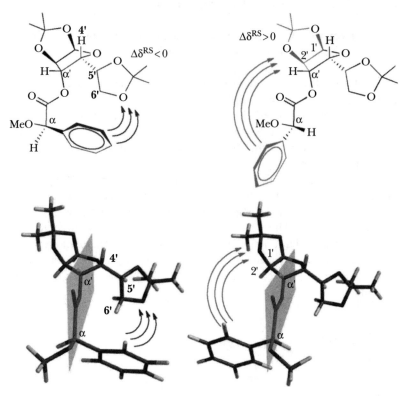

*Figure 3.4.* Position of the diagnostic hydrogens with respect to the phenyl group in the *sp* conformers of the two MPA esters of diacetone D-glucose.

Explanations for the different chemical shifts observed in the spectra of the (R)-MPA and the (S)-MPA derivatives are illustrated in Figure 3.4, which shows the position of the different protons with respect to the phenyl group in the *sp* conformer of the two derivatives. Thus, the H(4′)/H(5′)/H(6′) are under the shielding cone of the phenyl in the (R)-MPA ester (Figure 3.4a), while the H(1′)/H(2′) are under the shielding cone of the phenyl in the (S)-MPA ester (Figure 3.4b). In both cases, the diagnostic protons lie on the same side as the phenyl group in the corresponding *sp* conformer.

### 3.1.3. Example 2: Assignment of the Absolute Configuration of (−)-Isopulegol Using 9-AMA

The use of 9-AMA or any of the other AMAAs shown in Figure 1.3 entails the same distribution of sign/spatial location as with MPA, the only practical difference being that the intensity of the shielding is higher, and thus so are the values of $\Delta\delta^{RS}$.

The step-by-step procedure for the absolute configurational assignment of an unknown enantiomer of isopulegol using 9-AMA as the CDA is illustrated in Figure 3.5. In this case, the signals for diagnosis are H(5′)/H(6′)/H(10′) at one side and H(2′)/H(3′)/H(8′)/H(9′) at the other. The spectra show big differences in the chemical shifts between the two derivatives leading to the different values of $\Delta\delta^{RS}$ presented in Figure 3.5c [negative for H(2′)/H(3′)/H(8′)/H(9′) and positive for H(5′)/H(6′)/H(10′)]. In accordance with the sign/spatial location correlation for alcohols, protons H(2′)/H(3′)/H(8′)/H(9′) should be in front ($L_1$ in Figure 3.5d) and H(5′)/H(6′)/H(10′) in back of the tetrahedron around C$\alpha$′ ($L_2$ in Figure 3.5d), leading to the stereochemistry shown, which is coincident with that of (−)-isopulegol, that was the initial sample of known configuration.

As before, the differences observed in the chemical shifts in the two derivatives can be clearly understood when viewing Dreiding stereomodels that represent the sp conformation of the (R)- and (S)-9AMA esters; special attention should be paid to the relative position (with respect to the anthryl group) of the diagnostic protons of isopulegol (Figure 3.6).

In addition to the $\Delta\delta^{RS}$ signs for the proton signals, the assignment can also be carried out, or complemented, with the corresponding $^{13}$C $\Delta\delta^{RS}$ signs [72] (Section 1.5, Chapter 1). All that is necessary is to record the $^{13}$C-NMR spectra of the derivatives and apply to those $\Delta\delta^{RS}$ signs the same correlation model applied to protons [15]. Caution should be taken in this operation due to the small $^{13}$C shifts that in

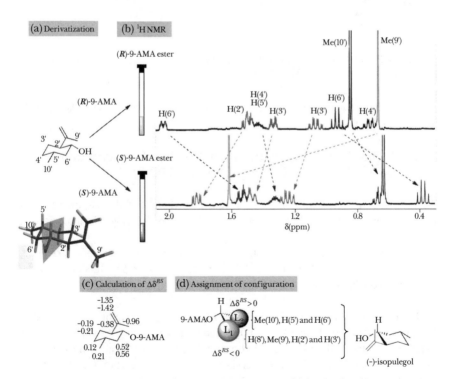

*Figure 3.5.* Main steps in the configurational assignment of (−)-isopulegol using 9-AMA.

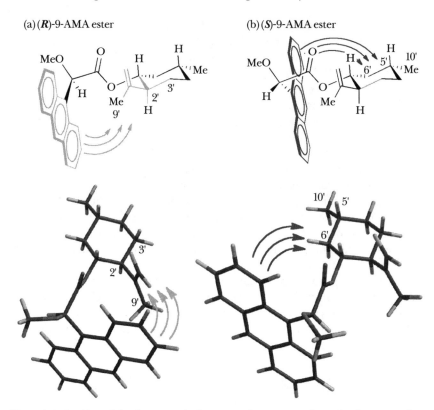

*Figure 3.6.* Position of the diagnostic hydrogens with respect to the anthryl group in the *sp* conformers of the two 9-AMA esters of (−)-isopulegol.

some cases can be close to the experimental error and therefore useless for the assignment [82].

### 3.1.4. Example 3: Assignment of the Absolute Configuration of (*R*)-Butan-2-ol Using MPA and $^{13}$C-NMR

The step-by-step procedure for the absolute configurational assignment of an unknown enantiomer of butan-2-ol using MPA and $^{13}$C NMR is illustrated in Figure 3.7.

A sample of a butan-2-ol of unknown configuration is separately derivatized with (*R*)- and (*S*)-MPA, and the $^{13}$C-NMR spectra of the MPA esters are recorded and assigned (Figures 3.7a and b). In this case, the signals for diagnosis are C(1′) on one side, and C(3′)/C(4′) on the other.

The calculated $\Delta\delta^{RS}$ data are shown in Figure 3.7c, displaying negative values for C(3′)/C(4′) and positive for C(1′). Thus, in accordance with the sign/spatial location correlation for secondary alcohols, C(3′)/C(4′) are located in the $L_1$ position, and C(1′) is in the $L_2$ position, as shown in Figure 3.7d. Therefore, the stereochemistry corresponds to (*R*)-butan-2-ol.

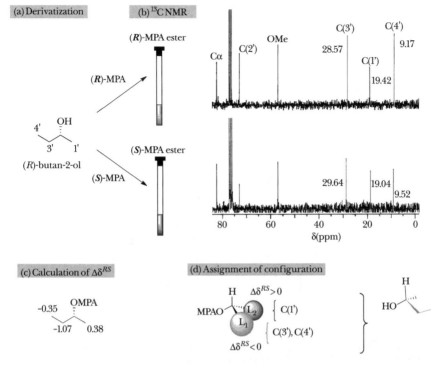

*Figure 3.7.* Main steps in the configurational assignment of (R)-butan-2-ol using MPA and $^{13}$C NMR.

Explanations for the different chemical shifts observed in the spectra of the (R)-MPA and (S)-MPA derivatives are illustrated in the Figure 3.8. In the *sp* conformer of the (R)-MPA ester, carbons C(3') and C(4') are located under the shielding cone of the phenyl group (Figure 3.8a), while carbon C(1') is under the shielding cone of the phenyl in the (S)-MPA ester (Figure 3.8b). Thus, C(3') and C(4') are more shielded in the (R)- than in the (S)-MPA ester (i.e., negative $\Delta\delta^{RS}$), and C(1') is more shielded in the (S)- than in the (R)-MPA ester (i.e., positive $\Delta\delta^{RS}$).

### 3.1.5. Simultaneous Derivatization of the Substrate with the (R)- and (S)-CDAs

When the structure of the substrate is simple enough to produce a NMR spectrum with just a few, well-separated signals, it is possible to simplify the procedure by producing the two derivatives simultaneously [40]. This means that the substrate is derivatized with a mixture of the (R)- and (S)-enantiomers of the CDA in a clearly different ratio (e.g., 2:1 or 3:1), producing a mixture of the (R)- and (S)-derivatives, whose spectra are recorded simultaneously in a single graph; this should allow identifying the signals due to each derivative. In this way, all the $\Delta\delta^{RS}$ can be measured from a single recording, that is, the assignments of the components of the mixture are carried out after only one derivatization and analysis of the spectrum.

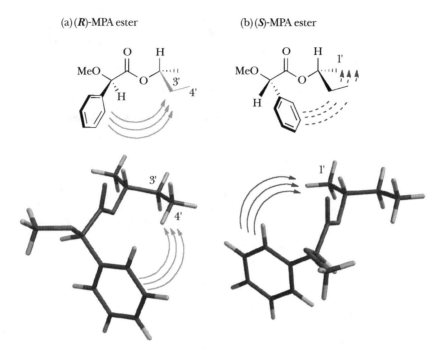

*Figure 3.8.* Position of the diagnostic carbons with respect to the phenyl group in the *sp* conformers of the two MPA esters of (*R*)-butan-2-ol.

Application of this procedure to a pure enantiomer of butan-2-ol with MPA is described in Section 3.1.6.

### 3.1.6. Example 4: Assignment of the Absolute Configuration of (*S*)-Butan-2-ol Using a 1:2 Mixture of (*R*)- and (*S*)-MPA

A sample of an unknown enantiomer of butan-2-ol is derivatized as usual with a 1:2 mixture of (*R*)- and (*S*)-MPA (Figure 3.9a). The resulting $^1$H-NMR spectrum shows well-separated signals for all the protons in the (*R*)- and (*S*)-MPA derivatives (Figure 3.9b). The diagnostic signals are Me(1′) on one side and the ethyl chain [CH$_2$(3′)/Me(4′)] on the other, and those from the (*R*)-MPA ester are clearly distinguished from those of the (*S*)-MPA ester by their different intensity. Calculation of $\Delta\delta^{RS}$ directly from the spectrum leads to the data shown in Figure 3.9c, and application of the sign/spatial location correlation for alcohols places the ethyl chain in the L$_2$ location (Figure 3.9d) and the Me(1′) in the L$_1$ location (Figure 3.9d). The stereochemistry shown corresponds to (*S*)-butan-2-ol, coincident with that of the initial sample.

The behavior observed in the $^1$H-NMR spectrum is easily explained by examination of the conformers shown in Figures 3.9e and f. Me(1′) presents a negative $\Delta\delta^{RS}$, because it is more shielded in the *sp* conformer of the (*R*)- than in that of the (*S*)-MPA ester (Figure 3.9e). On the other hand, positive $\Delta\delta^{RS}$ values are obtained

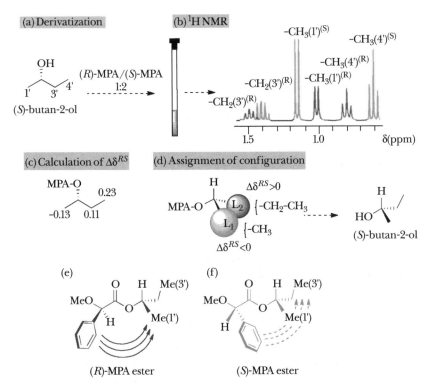

Figure 3.9. Main steps in the configurational assignment of (S)-butan-2-ol using 1:2 mixture of (R)- and (S)-MPA.

for Me(4′) and CH$_2$(3′), because they are more shielded in the *sp* conformer of the (S)- than in that of the (R)-MPA ester.

### 3.1.7. Example 5: Assignment of the Absolute Configuration of (−)-Menthol Using a 2:1 Mixture of (R)- and (S)-9-AMA and $^{13}$C NMR

A sample of menthol of unknown stereochemistry is derivatized with a 2:1 mixture of (R)- and (S)-9-AMA (Figure 3.10a). The resulting $^{13}$C-NMR spectrum shows well-separated signals for all the carbons in the (R)- and (S)-9-AMA esters (Figure 3.10b). NMR signals corresponding to the (R)-9-AMA ester are clearly distinguished from those from the (S)-9-AMA ester by their different intensity [those of the (R) are more intense than those of the (S)]. The diagnostic signals are those corresponding to C(2′)/C(3′)/C(7′)/C(8′)/C(9′)/C(5′)/C(6′)/C(10′).

Calculation of the $\Delta\delta^{RS}$ directly from the spectrum leads to the data shown in Figure 3.10c: positive for C(5′)/C(6′)/C(10′) and negative for C(3′)/C(7′)/C(8′)/C(9′). Application of the sign/spatial location correlation for secondary alcohols places C(3′)/C(7′)/C(8′)/C(9′) in the L$_1$ position in the structure shown

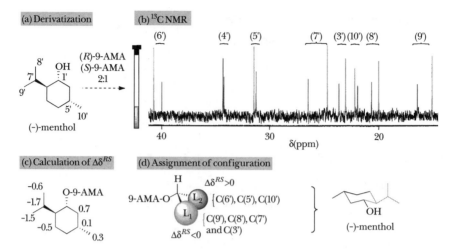

Figure 3.10. Main steps in the configurational assignment of (−)-menthol using a 2:1 mixture of (R)- and (S)-9-AMA and $^{13}$C NMR.

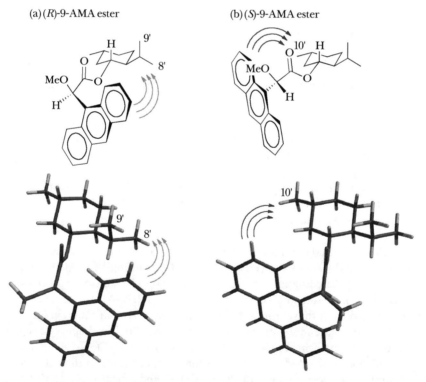

Figure 3.11. Position of the diagnostic carbons with respect to the anthryl group in the sp conformers of the two MPA esters of (−)-menthol.

in Figure 3.10d, and C(5′)/C(6′)/C(10′) in the $L_2$ position of the same structure. Therefore, the stereochemistry shown corresponds to (−)-menthol.

Carbons C(3′)/C(7′)/C(8′)/C(9′) present negative $\Delta\delta^{RS}$ values because they are more shielded by the anthryl group in the *sp* conformer of the (*R*)-9-AMA ester than in that of the (*S*)-9-AMA ester, as shown in Figure 3.11. On the other hand, carbons C(5′)/C(6′)/C(10′) are more shielded in the *sp* conformer of the (*S*)-9-AMA ester (Figure 3.11).

## 3.1.8. MTPA as the CDA for Secondary Alcohols

The use of MTPA as the CDA is similar to that of MPA from the experimental point of view [37, 38]. However, an important difference is found in the conformational equilibrium of the MTPA esters [38], and it results in a different correlation between the signs and the spatial locations of $L_1$ and $L_2$.

Thus, theoretical and experimental data have shown that the MTPA esters of a secondary alcohol are composed by three main conformers [38] (*sp1, sp2* and *ap1*) instead of the two (*sp, ap*) found in the AMAA esters [37]. The main conformation is *ap1*, and this can be used to interpret the chemical shifts in much the same way as we resort to the *sp* conformer in the MPA esters. It is necessary to point out that in the *ap1* conformations, deshielding effects predominate (unlike the shielding effects that occur with MPA and the other AMAAs). Figure 3.12 shows the three conformers in the (*R*)- and (*S*)-MTPA esters of a secondary alcohol, as well as the shielding/deshielding effects and the resulting correlation between the $\Delta\delta^{SR}$ signs and the absolute stereochemistry. The substituent with a negative $\Delta\delta^{SR}$ is located in front of the tetrahedron around the asymmetric carbon ($L_1$), while the substituent with the positive $\Delta\delta^{SR}$ is located at the back ($L_2$).

For historical reasons, the comparison between the spectra of the two MTPA derivatives is carried out by a subtraction defined as the chemical shifts in the (*S*)-MTPA derivative minus those of the (*R*)-MTPA derivative (this is the opposite of what is done with the other CDAs); this leads to differences in $\Delta\delta^{SR}$ instead of in $\Delta\delta^{RS}$ [14].

## 3.1.9. Example 6: Assignment of the Absolute Configuration of (−)-Borneol Using MTPA

A sample of a pure enantiomer of borneol [the (−)-borneol isomer] is separately derivatized with (*R*)- and (*S*)-MTPA, affording the corresponding MTPA esters (Figure 3.13).

The diagnostic protons, located at the two sides of the plane in the *ap* conformation of the MTPA ester, are H(4′)/H(5′)/H(6′) on one side of the chiral center and H(3′)/H(10′) on the other. Partial $^1$H-NMR spectra (Figure 3.13b) with the signals for those protons show that H(4′)/H(5′)/H(6′) are more shielded in the (*R*)- than in the (*S*)-MTPA ester (positive $\Delta\delta^{SR}$, Figure 3.13c), while the signals for H(3′)/H(10′) are more shielded in the (*S*)- than in the (*R*)-MTPA ester (negative $\Delta\delta^{SR}$, Figure 3.13c). Comparison of this sign distribution with the correlation model for MTPA esters of secondary alcohols (Figure 3.13d) locates

**48** ■ The Assignment of the Absolute Configuration by NMR

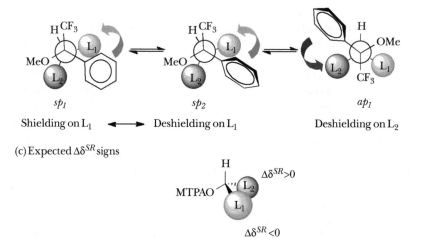

Figure 3.12. Shielding/deshielding effects in the main conformers of MTPA esters, and the expected $\Delta\delta^{SR}$ signs.

H(4′)/H(5′)/H(6′) at the $L_2$ location in the structure shown in Figure 3.13d, and H(3′)/H(10′) at the $L_1$ location.

Rationalization of this correlation between $\Delta\delta^{SR}$ signs and spatial location is obtained by examination of the Dreiding stereomodels representing the *ap1* conformations and the corresponding deshielding effects (Figure 3.14).

## 3.1.10. Summary

As we have shown in the preceding sections (Sections 3.1.1 to 3.1.9), the absolute configuration of secondary alcohols can be determined by comparison of the NMR spectra of the (R)- and (S)-derivatives of a number of different AMAA acids [14, 15, 35–40] (AMAAs), the most common one being MPA [38]. It is commercially available and provides $\Delta\delta^{RS}$ values that, in most cases, are large enough to be significant and have positive or negative signs homogeneously distributed in $L_1$ and $L_2$, allowing for confidence in the assignments.

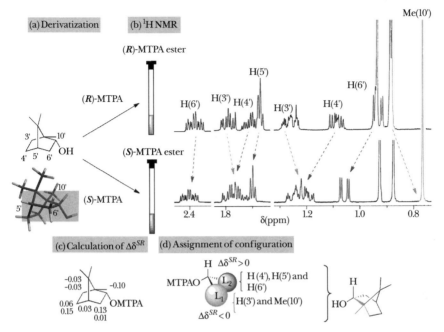

*Figure 3.13.* Main steps in the configurational assignment of (−)-borneol using MTPA.

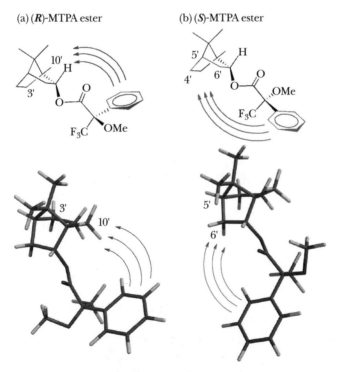

*Figure 3.14.* Position of the diagnostic hydrogens with respect to the phenyl group in the *ap1* conformers of the two MTPA esters of (−)-borneol. Curved arrows indicate the deshielding effects.

50 ■ The Assignment of the Absolute Configuration by NMR

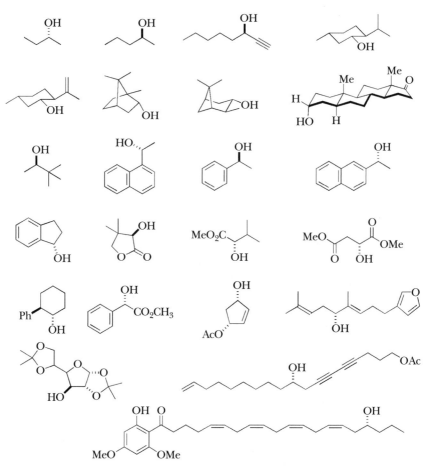

*Figure 3.15.* Selection of chiral secondary alcohols with known absolute configuration used to validate the procedures described in this chapter.

The substitution of the phenyl group by larger aromatic systems yields CDAs with higher shielding effects and particular conformational characteristics. Among them, 9-AMA is the most potent [37, 39], producing very high $\Delta\delta^{RS}$ values. Unfortunately, it is not commercially available (see Chapter 2 for its preparation), and because of this, it is usually reserved for those cases where MPA generates small $\Delta\delta^{RS}$ values due to the long distance from the CDA to the protons under consideration. Finally, MTPA, like MPA, is commercially available, but the presence of the very polar trifluoromethyl substituent imposes a conformational composition on the MTPA esters that is more complex than that of the MPA esters [38] (three instead of two main conformers), making the NMR of the MTPA derivatives sensitive to slight structural, conformational, or experimental changes and prone to produce small $\Delta\delta^{SR}$ values, inhomogeneous sign distributions, and misassignments [14, 82].

It is important to point out that, for historical reasons, when MTPA is used, the chemical shift differences between the MTPA ester derivatives are measured as $\Delta\delta^{SR}$ instead of $\Delta\delta^{RS}$, as is done with all the other CDAs [14].

Also, and in order to prevent confusion, it should be stressed that when preparing the MTPA esters, the use of MTPA or of its acid chloride results in different nomenclatures due to the order of priority of the substituents in MTPA and MTPA chloride [85]. Thus, while (R)-MTPA produces the (R)-MTPA ester, the (R) enantiomer of MTPA-Cl produces the (S)-MTPA ester, and the reverse is true for the (S)-enantiomers.

The double-derivatization assignment of secondary alcohols can also be carried out by $^{13}$C NMR of the same derivatives used for $^1$H NMR. The stereochemical correlation between the $\Delta\delta^{RS}$ values and the spatial locations of $L_1$ and $L_2$ are the same for both proton and carbon nuclei, allowing the recording of both types of NMR spectra from the same samples. In this way, $^1$H $\Delta\delta^{RS}$ and $^{13}$C $\Delta\delta^{RS}$ signs can be used together or separately for the assignment [72]. $^{13}$C $\Delta\delta^{RS}$ values are very small in comparison with the corresponding $^1$H data, but in most cases, the signs are sufficient for assignment.

Finally, as will be discussed in Chapter 4, the assignment of the absolute configuration of secondary alcohols can also be performed by $^1$H NMR using only one derivative, either the (R) or the (S) ester, with MPA or with 9-AMA as the CDA [41–43].

Figure 3.15 shows a selection of secondary alcohols with known absolute configuration and mixed structures; these have been used to validate the procedures described in this chapter [35–43]. Other examples of applications of this methodology can be found in the literature [97–142].

## 3.2. β-CHIRAL PRIMARY ALCOHOLS

Primary alcohols, with a chiral carbon at the β position, can be assigned within certain limitations as 9-AMA ester derivatives [14, 15, 44–46]. Depending on the nature of the $L_1/L_2$ substituents, there are two different $\Delta\delta^{RS}$ sign/spatial location correlations [45, 46]. Figure 3.16 shows those correlations, as well as the essentials of the procedure.

### 3.2.1. Assignment of β-Chiral Primary Alcohols as 9-AMA Esters

In the context of NMR assignment, the basic difference between secondary and primary alcohols reside in the presence of an additional C(1′)-C(2′) bond between the alcohol and the chiral carbon [45]. This means that the distance between the aryl ring of the CDA and the $L_1/L_2$ groups is longer, and therefore the intensity of the shieldings is lower. This problem can be solved by resorting to a CDA with an anthryl group, such as 9-AMA, instead of a phenyl group, as in MPA. Unfortunately, rotation around the C(1′)-C(2′) bond takes place very easily, and therefore 9-AMA esters, in which the $L_1/L_2$ groups show low polarity, are composed of several conformers in close equilibrium [45]. Figure 3.17 illustrates the three main conformations that define the (R)-9-AMA esters of a primary alcohol (Figure 3.17a). In most cases, the NMR spectra and the shieldings observed can be explained assuming that the representative conformation is *sp-a/a*. This

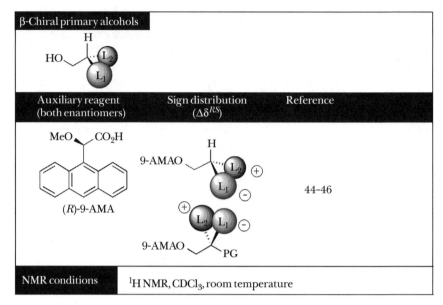

*Figure 3.16.* The assignment of β-chiral primary alcohols at a glance: CDA, sign distributions ($\Delta\delta^{RS}$), experimental conditions, and references.

conformer leads to the shieldings on $L_1/L_2$ indicated in Figures 3.17b and c, and to the sign/spatial location correlation shown: the substituent located at the same side of the anthryl group in the *sp-a/a* conformation of the (*R*)-9-AMA ester (i.e., $L_1$) has a negative $\Delta\delta^{RS}$ and is located in front of the tetrahedron around the asymmetric carbon, while the other substituent (positive $\Delta\delta^{RS}$, i.e., $L_2$) is placed in the back. The $\Delta\delta^{RS}$ values are clearly lower than those with the secondary alcohols; they are sometimes so close to the experimental error that no safe assignment can be deduced.

The small shifts observed for primary alcohols prevent the use of $^{13}$C NMR for assignment [72].

### 3.2.2. Example 7: Assignment of the Absolute Configuration of (*S*)-2-Methylbutan-1-ol Using 9-AMA

A sample of a pure enantiomer of 2-methylbutan-1-ol, known to be the (*S*)-isomer, was separately derivatized with (*R*)- and (*S*)-9-AMA. Figure 3.18 shows the partial spectra of the derivatives containing the signals due to diagnostic protons located at the two sides of the plane and defined by the *sp-a/a* conformation: H(3′)/H(4′) on one side and H(5′) on the other.

When comparing the (*R*)- with the (*S*)-9-AMA derivative, the signals for H(3′)/H(4′) move to a higher field (positive $\Delta\delta^{RS}$; Figure 3.18c), while H(5′) moves to a lower field (negative $\Delta\delta^{RS}$; Figure 3.18c). Application of the correlation model (Figures 3.17 and 3.18d) places H(5′) in front of the tetrahedron around the asymmetric carbon (i.e., $L_1$) and H(3′)/H(4′) at the back (i.e., $L_2$); this results in the stereochemistry shown in Figure 3.18d.

(a) (*R*)-9-AMA ester

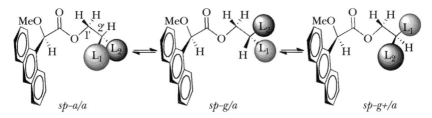

(b) (*R*)-9-AMA ester (*sp-a/a* conformer)

(c) (*S*)-9-AMA ester (*sp-a/a* conformer)

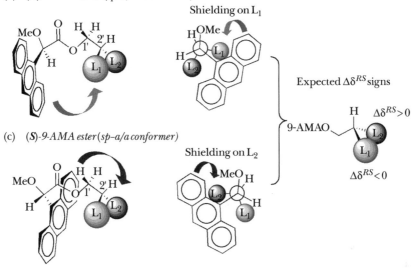

*Figure 3.17.* Main conformations of the 9-AMA esters of primary alcohols. Shielding effects in the most representative conformers and expected $\Delta\delta^{RS}$ signs are shown.

Rationalization of the selective shieldings and shifts observed for each derivative and of the correlation between the $\Delta\delta^{RS}$ sign and spatial location is obtained by analysis of Dreiding models representing the *sp-a/a* conformation (Figure 3.19).

### 3.2.3. Absolute Configuration of Primary Alcohols with Polar Groups as 9-AMA Esters

As indicated in Section 3.2.1, the 9-AMA ester derivatives of primary alcohols are conformationally complex, and in fact, the NMR method just shown for primary alcohols [45] should not be employed with compounds where either one of the $L_1/L_2$ substituents does not have any hydrogens, or when those present are far away from the functional group [45, 46].

A method using 9-AMA as the CDA was designed for the assignment of the absolute configuration of β-chiral primary alcohols bearing polar groups (indicated by Pg in figures) as substituents [46]. Usually, those polar groups are devoid

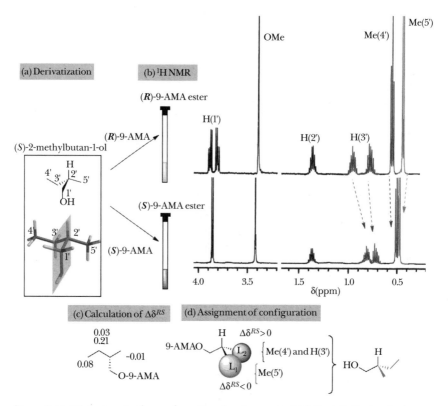

*Figure 3.18.* Main steps in the configurational assignment of (S)-2-methylbutan-1-ol using 9-AMA.

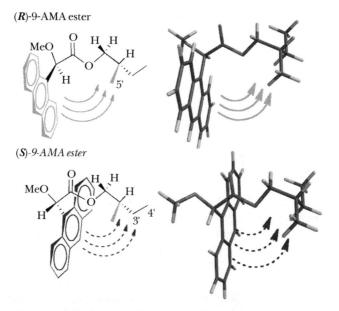

*Figure 3.19.* Position of the diagnostic hydrogens with respect to the anthryl group in the *sp-a/a* conformers of the two 9-AMA esters of (S)-2-methylbutan-1-ol.

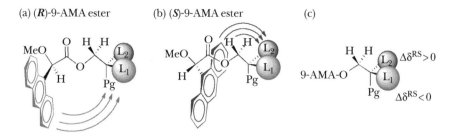

Pg = polar group (OH, OR, NHC(O)R, NHC(O)OR, CH$_2$X...)
L$_1$ or L$_2$ = H

*Figure 3.20.* Main conformations of the 9-AMA esters of primary alcohols bearing polar groups. Shielding effects in the most representative conformers and expected $\Delta\delta^{RS}$ signs are shown. In the models, either L$_1$ or L$_2$ is a hydrogen atom.

of protons that are useful for making assignments; some examples include OH, OR, Cl, NHC(O)R, NHC(O)OR, C(O)OR, and CH$_2$Br.

The presence of a polar group is sufficient to modify the conformational equilibrium of a 9-AMA ester, and the most representative conformer is now an *sp* one [46]. In this way, the *sp* conformer can be used to explain the $^1$H-NMR spectra and the observed shieldings (Figures 3.20a and b), leading to the correlation model indicated in Figure 3.20c. In this case, the diagnostic signals correspond to the other substituent and to the hydrogen that is directly bonded to the asymmetric C(2′).

### 3.2.4. Example 8: Assignment of the Absolute Configuration of (S)-2-Chloropropan-1-ol Using 9-AMA

A sample of a pure (S)-2-chloropropan-1-ol was separately derivatized with (R)- and with (S)-9-AMA. Figure 3.21 shows the partial $^1$H-NMR spectra of the esters containing the signals for the diagnostic protons located on the two sides of the plane defined by the *sp* conformation: Me(3′) on one side and H(2′) on the other.

When going from the (R)- to the (S)-9-AMA derivatives, the signal of H(2′) moves to a lower field (negative $\Delta\delta^{RS}$), while Me(3′) moves to a higher field (positive $\Delta\delta^{RS}$). Application of the correlation model places Me(3′) behind and H(2′) in front of the tetrahedron around the asymmetric center; this leads to the stereochemistry shown in Figure 3.21d.

Rationalization of the shieldings and chemical shifts observed in each derivative, together with an explanation of the correlation between the $\Delta\delta^{RS}$ sign and the spatial location, can be easily obtained by examination of the representations of the *sp* conformations shown in Figure 3.22.

In the most representative NMR conformation, H(2′) is more shielded in the (R)- than in the (S)-9-AMA ester, thus a negative $\Delta\delta^{RS}$ is obtained. On the other hand, Me(3′) is more shielded in the (S)- than in the (R)-9-AMA ester, and therefore a positive $\Delta\delta^{RS}$ is observed.

# 56 ■ The Assignment of the Absolute Configuration by NMR

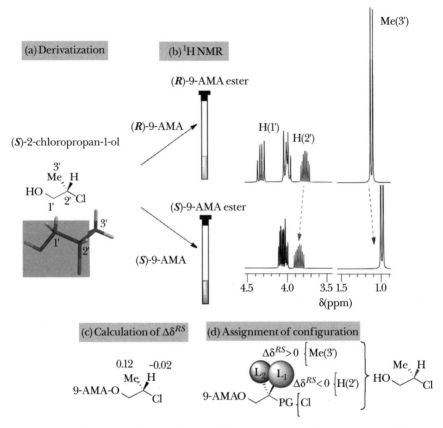

*Figure 3.21.* Main steps in the configurational assignment of (S)-2-chloropropan-1-ol using 9-AMA.

## 3.2.5. Summary

To summarize, the absolute configuration of primary alcohols can be determined in many cases by comparison of the $^1$H-NMR spectra of the (R)- and (S)-9-AMA derivatives [44–46]. Even with that reagent, the shifts are much lower than those obtained with secondary alcohols. This sometimes leads to $\Delta\delta^{RS}$ values so close to the experimental error that no safe assignment can be derived [82]. In addition, the conformational compositions of these 9-AMA ester derivatives do not clearly show a single main conformer, as is seen with secondary alcohols; this sometimes leads to uncertainty. As a solution to this, it has been proposed that the structures of those primary alcohols that could lead to misassignments should be identified, but this involves ab initio calculations that are beyond the scope of this book [45].

These considerations indicate that the assignment of primary alcohols by NMR does not have the same guarantee as with the other functional groups, and it should be used with caution. This and the small shifts suggest that $^{13}$C-NMR data are not sufficient for assignments [72].

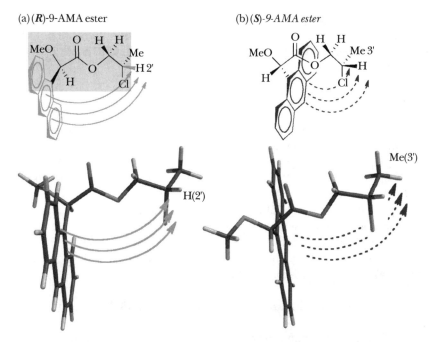

*Figure 3.22.* Position of the diagnostic hydrogens with respect to the anthryl group in the *sp* conformers of the two 9-AMA esters of (S)-2-chloropropan-1-ol.

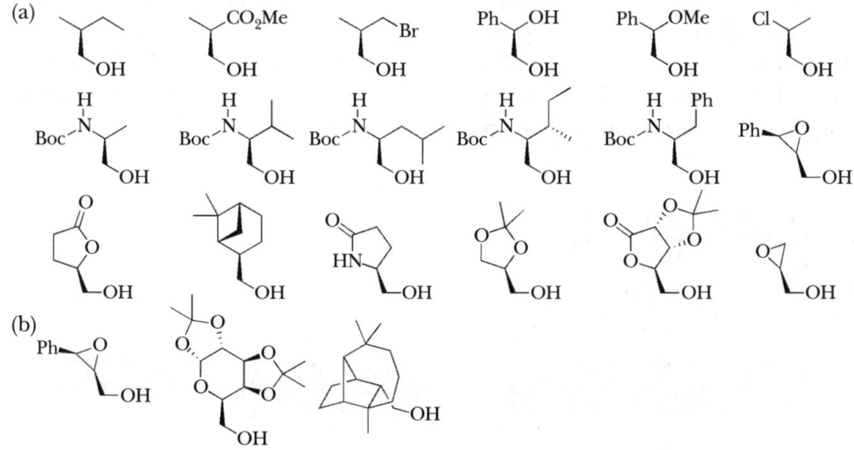

*Figure 3.23.* Selection of β-chiral primary alcohols with known absolute configuration used to validate the procedures described in this chapter.

Nevertheless, when one of the two substituents at the asymmetric carbon is highly polar, the conformational equilibrium is clearly in favor of the *sp* conformation, and the 9-AMA esters allow a safe configurational assignment.

Figure 3.23a shows a selection of primary alcohols with known absolute configurations that have been used to validate the procedure described in this

**58** ■ The Assignment of the Absolute Configuration by NMR

chapter [44–46]. Figure 3.23b shows three examples that are described in the literature but that do not follow the sign/spatial location correlation. Other examples of applications of this methodology can be found in the literature [86, 143–146].

## ■ 3.3. ALDEHYDE CYANOHYDRINS

The assignment of the absolute configuration of aldehyde and ketone cyanohydrins by NMR is resolved by considering cyanohydrins to be a special class of alcohols [47–49]. In this way, the OH group is used for derivatization with a CDA bearing a carboxylic acid; MPA is the reagent of choice. When examining the structure of the MPA ester derivatives of aldehyde cyanohydrins, it can be immediately seen that one of the substituents has no protons (i.e., the CN group); therefore, the assignment is forced to use a combination of $^1$H and $^{13}$C NMR [72].

Figure 3.24 shows a summary of the structures of aldehyde cyanohydrins, that of the auxiliary reagent, and the correlation between the $\Delta\delta^{RS}$ sign distribution and the spatial location of the substituents.

### 3.3.1. Assignment of Aldehyde Cyanohydrins as MPA Esters

The conformational composition of the MPA esters of aldehyde cyanohydrins [47] resembles that of the secondary alcohols [37], being represented by a rapid equilibrium between *sp* (the most populated one) and *ap* conformations. Figure 3.25 shows the (*R*)- and the (*S*)-MPA derivatives of an aldehyde cyanohydrin. As in the secondary alcohols, the phenyl group of MPA produces a shielding that selectively affects the substituent located on the same side of the plane defined by the *sp* conformation (see Figures 3.25a and b). When those shieldings are analyzed and the two derivatives are compared, the corresponding $\Delta\delta^{RS}$ signs ($^1$H/$^{13}$C for L and $^{13}$C

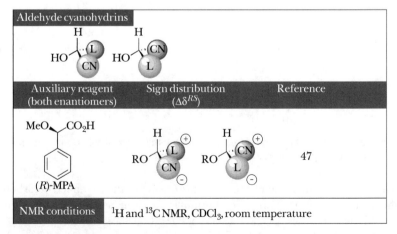

*Figure 3.24.* The assignment of aldehyde cyanohydrins at a glance: CDA, sign distributions ($\Delta\delta^{RS}$), experimental conditions, and references.

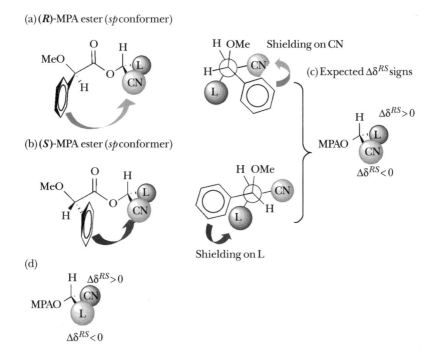

*Figure 3.25.* Shielding effects in the most representative conformers of MPA esters of aldehyde cyanohydrins and expected $\Delta\delta^{RS}$ signs.

for CN) are derived, and the correlation between those signs and the spatial locations of L and CN are established [47] (Figures 3.25c and d).

The correlation between the $\Delta\delta^{RS}$ signs and the absolute configuration of the aldehyde cyanohydrin can be deduced easily by analyzing the NMR-representative conformers shown in Figures 3.25a and b. Thus, taking as the model compound the aldehyde cyanohydrin shown in those Figures, the L group (both protons and carbons) is more shielded in the (S)- than in the (R)-MPA derivative (Figure 3.25b), while the carbon in the CN group is more shielded in the (R)- than in the (S)-MPA derivative (Figure 3.25a). Therefore, a negative $\Delta\delta^{RS}$ is expected for CN and a positive one for L [47] (Figure 3.25c). The opposite behavior is found for the enantiomeric cyanohydrin (Figure 3.25d).

### 3.3.2. Example 9: Assignment of the Absolute Configuration of (*R*)-2-Hydroxy-3-Methylbutanenitrile Using MPA

A sample of the cyanohydrin of isobutylaldehyde (Figure 3.26) was derivatized with (R)- and (S)-MPA, producing the corresponding MPA esters (Figure 3.26a). The diagnostic protons and carbons are those located on the two sides of the plane defined by the *sp* conformation [i.e., the isopropyl group, H(3′)/Me(4′)/Me(5′)] on one side and the CN group (its carbon signal) on the other.

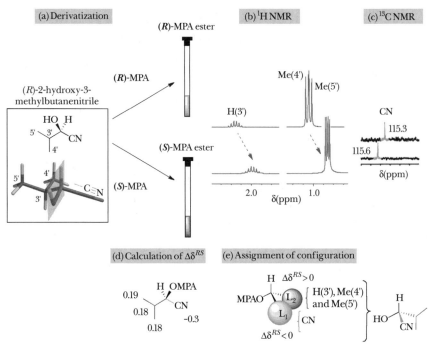

*Figure 3.26.* Main steps in the configurational assignment of (R)-2-hydroxy-3-methylbutanenitrile using MPA.

Comparison of the ¹H-NMR spectra (Figure 3.26b) indicates that the signals for H(3′)/Me(4′)/Me(5′) are more shielded in the (S)-MPA derivative (i.e., positive $\Delta\delta^{RS}$; Figure 3.26d), while the corresponding ¹³C-NMR spectra (Figure 3.26c) shows that the CN carbon is more shielded in the (R)-MPA derivative (i.e., negative $\Delta\delta^{RS}$, Figure 3.26d). Application of the correlation model (Figures 3.25 and 3.26e) places the CN group in the $L_1$ location. On the other hand, H(3′)/Me(4′)/Me(5′) occupy the $L_2$ location, leading to the stereochemistry shown in Figure 3.26e, which corresponds to the (R) enantiomer [i.e., (R)-2-hydroxy-3-methylbutanenitrile].

Explanation of the shifts observed is easily obtained by building the corresponding Dreiding models in the *sp* conformation (Figure 3.27), bisecting the structures by the *sp* plane, and observing the positions of the cyanohydrin substituents relative to the phenyl group of MPA.

### 3.3.3. Example 10: Assignment of the Absolute Configuration of (R)-2-Hydroxy-2-(4-Methoxyphenyl)Acetonitrile Using MPA and ¹³C NMR

Following the steps already mentioned in the previous examples, a sample of a single enantiomer of 2-hydroxy-2-(4-methoxyphenyl)acetonitrile of unknown configuration is separately derivatized with (R)- and (S)-MPA (step 1). The

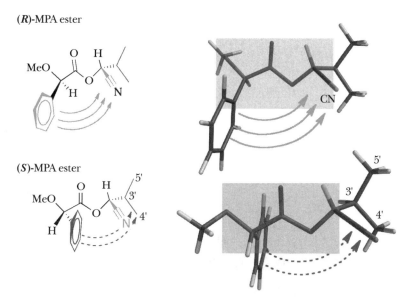

*Figure 3.27.* Position of the diagnostic hydrogens and carbons with respect to the phenyl group in the *sp* conformers of the two MPA esters of (R)-2-hydroxy-3-methylbutanenitrile.

$^{13}$C-NMR spectra of the derivatives are then registered, the signals from the carbons of the substituents are identified (Figures 3.28a and b), and their chemical shifts are compared in order to obtain the corresponding $\Delta\delta^{RS}$. The differences, although small, are significant (Figure 3.28c) and show a negative sign for the CN carbon ($\Delta\delta^{RS} = -0.3$), that is more shielded in the (R)-MPA ester. The aromatic carbons show positive signs ($\Delta\delta^{RS} = +0.10$ and $+0.40$), so they are more shielded in the (S)-MPA ester and can be identified unambiguously as belonging to the other substituent, C(3')/C(4'). The $\Delta\delta^{RS}$ value obtained for the distant methoxy group is too close to zero to provide a significant sign and therefore is not useful for the assignment (Figure 3.28c).

The absolute configuration is derived from the comparison of the graphical model illustrating the correlation between the spatial location of the substituents and the experimental $\Delta\delta^{RS}$ signs (Figure 3.28d; the same stereomodel used with $^1$H NMR). In this case, the comparison leads to the placement of the cyano group (negative $\Delta\delta^{RS}$) in the $L_1$ location and the aryl substituent (positive $\Delta\delta^{RS}$) in the $L_2$ location, leading to the (R) absolute configuration shown in Figure 3.28d.

Dreiding models representing the main conformations (i.e., *sp*) of the (R)- and (S)-MPA derivatives show how the bisecting plane of the *sp* conformer distributes the two substituents of the cyanohydrin (CN and aryl groups) at different positions with respect to the MPA phenyl group, and they explain the shieldings observed in each derivative (Figure 3.29).

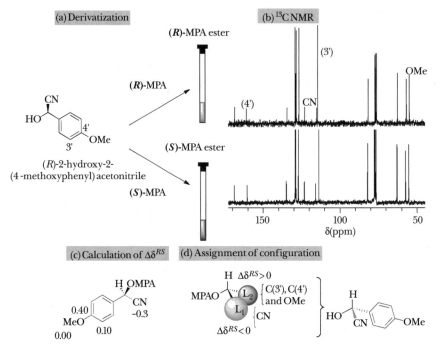

*Figure 3.28.* Main steps in the configurational assignment of (R)-2-hydroxy-2-(4-methoxyphenyl)acetonitrile using MPA and $^{13}$C NMR.

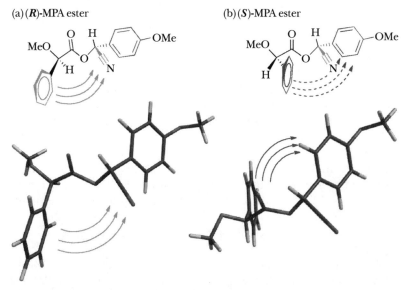

*Figure 3.29.* Position of the diagnostic carbons with respect to the phenyl group in the *sp* conformers of the two MPA esters of (R)-2-hydroxy-2-(4-methoxyphenyl)acetonitrile.

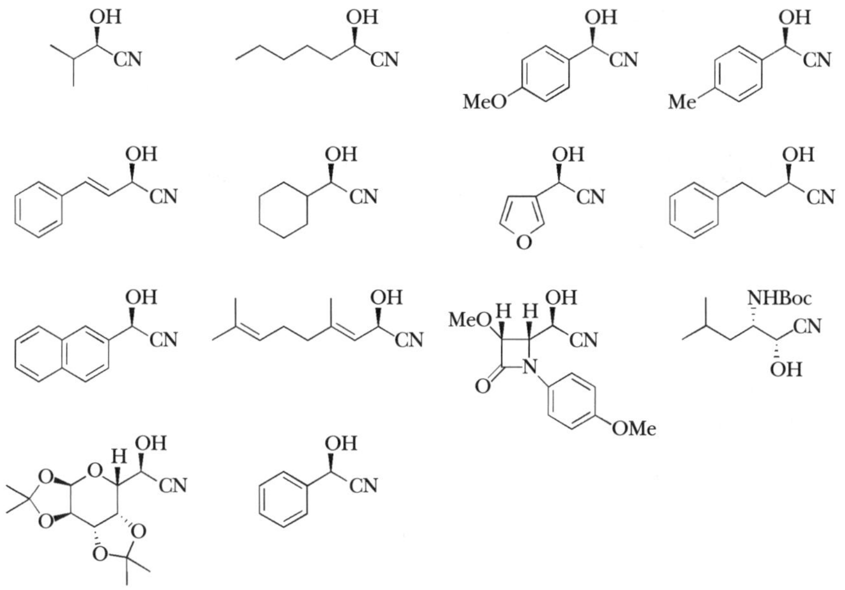

*Figure. 3.30.* Selection of aldehyde cyanohydrins with known absolute configuration used to validate the procedures described in this chapter.

### 3.3.4. Summary

The absolute configuration of aldehyde cyanohydrins is easily determined by double derivatization with (R)- and (S)-MPA and comparison of the NMR spectra of the derivatives [47]. The main conformation places the L and CN groups in positions that are adequate for generating the signals used in the diagnosis. Since the CN has no protons, the assignment is carried out using the $^1$H $\Delta\delta^{RS}$ sign for the L substituent and the $^{13}$C $\Delta\delta^{RS}$ sign for the CN substituent. Of course, the assignment can also be carried out using only $^{13}$C NMR for both the CN and the L substituents [47, 72].

Figure 3.30 shows a selection of cyanohydrins with known absolute configuration that have been used to validate this procedure [47]. Other examples of application of this methodology can be found in the literature [147].

## ■ 3.4. KETONE CYANOHYDRINS

Similarly to the aldehyde cyanohydrins [47], the absolute configuration of ketone cyanohydrins can be assigned by NMR using MPA as the auxiliary reagent, comparing the results to those of the corresponding MPA ester derivatives [48, 49]. Figure 3.31 presents the structures of the substrates and the auxiliary, the correlation between the $\Delta\delta^{RS}$ signs ($^1$H and $^{13}$C NMR), the absolute configuration, and the essentials of the experimental conditions.

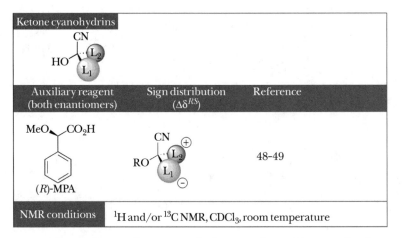

*Figure 3.31.* The assignment of ketone cyanohydrins at a glance: CDA, sign distribution ($\Delta\delta^{RS}$), experimental conditions, and references.

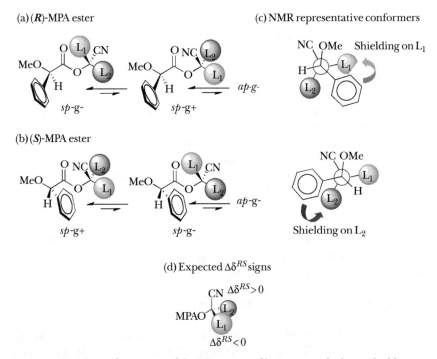

*Figure 3.32.* Main conformations of the MPA esters of ketone cyanohydrins. Shielding effects in the most representative conformers and expected $\Delta\delta^{RS}$ signs.

### 3.4.1. Assignment of Ketone Cyanohydrins as MPA Esters

From a structural point of view, MPA esters of ketone cyanohydrins [48, 49] resemble those of tertiary alcohols, but unfortunately, this is not of much help, because there is no established way to assign the configuration of chiral tertiary alcohols by NMR.

The conformational study of MPA esters of ketone cyanohydrins shows that the presence of the highly polar CN group is essential for determining the existence of main conformers. There are now three main conformers in fast equilibrium: two *sp* forms, which differ in their rotation around the O-Cα' bond, and one *ap* form; the two *sp* forms are the most abundant ones [48] (Figure 3.32).

When we examine the positions of the $L_1/L_2$ and CN substituents relative to the phenyl group of the MPA moiety, a selective and specific shielding is expected in the (*R*)- and in the (*S*)-MPA derivatives (Figures 3.32a and b). The *sp* conformations are the most representative ones for the interpretation of the spectra (Figure 3.32c), and they lead to a correlation between the $\Delta\delta^{RS}$ signs and the spatial location [48] of $L_1/L_2$ (Figure 3.32d).

Examples of applications of this methodology by $^1$H and $^{13}$C NMR follow in Sections 3.4.2 and 3.4.3.

### 3.4.2. Example 11: Assignment of the Absolute Configuration of (1*R*, 2*S*, 5*R*)-1-Hydroxy-2-Isopropyl-5-Methylcyclohexanecarbonitrile Using MPA

A sample of a pure enantiomer of the cyanohydrin of (2*S*,5*R*)-2-isopropyl-5-methylcyclohexanone (known to be the 1*R*,2*S*,5*R* isomer; Figure 3.33) was derivatized with (*R*)- and (*S*)-MPA. The *sp* plane divides the molecule into two parts that serve to define the protons/carbons useful for assignment (Figure 3.33a): the isopropyl group [protons H(7')/Me(8')/Me(9')] on one side of the plane and the protons H(6')/Me(10') on the other. The spectra of the two MPA esters are presented in Figure 3.33b, and a comparison shows that the signals for the protons H(7')/Me(8')/Me(9') are more shielded in the (*R*)-MPA derivative (negative $\Delta\delta^{RS}$), while the H(6')/Me(10') are more shielded in the (*S*)-MPA derivative (positive $\Delta\delta^{RS}$; Figure 3.33c). Comparison of these results with the correlation model (Figure 3.33d) show that the H(7')/Me(8')/Me(9') protons are in the $L_1$ location in the structure shown in Figure 3.33d, and the H(6')/Me(10') protons are at the rear (i.e., $L_2$; Figure 3.33d); this leads to the stereochemistry shown in Figure 3.33d, which corresponds to the (1*R*) isomer: (1*R*, 2*S*, 5*R*)-1-hydroxy-2-isopropyl-5-methylcyclohexanecarbonitrile.

Rationalization of these results and an explanation of the spectra and shifts can be easily obtained by examination of Dreiding models of the (*R*) and (*S*)-MPA esters in the *sp* conformation (Figure 3.34); bisect the structures through the *sp* plane and observe the spatial position of the cyanohydrin substituents relative to the phenyl group of MPA.

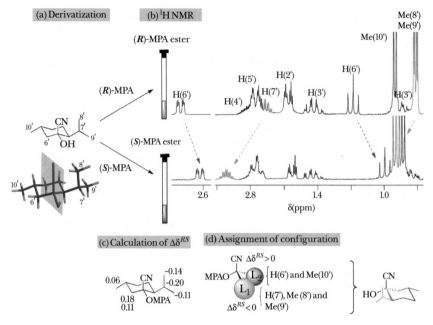

*Figure 3.33.* Main steps in the configurational assignment of (1R, 2S, 5R)-1-hydroxy-2-isopropyl-5-methylcyclohexanecarbonitrile using MPA.

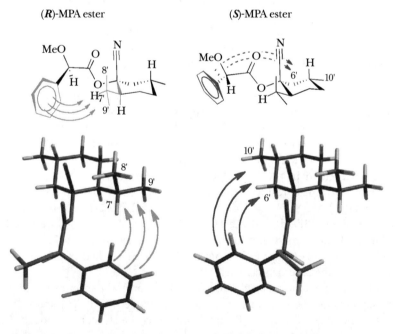

*Figure 3.34.* Position of the diagnostic hydrogens with regard to the phenyl group in the *sp* conformers of the two MPA esters of (1R, 2S, 5R)-1-hydroxy-2-isopropyl-5-methylcyclohexanecarbonitrile.

### 3.4.3. Example 12: Assignment of the Absolute Configuration of (S)-2-Hydroxy-2,4-Dimethylpentanenitrile Using MPA and $^{13}$C NMR

As in the previous case, an unknown enantiomer of 2-hydroxy-2, 4-dimethylpentanenitrile is derivatized with (R)- and (S)-MPA, the spectra of the derivatives are registered, and the carbon signals originated by the methyl and isobutyl substituents are identified (Figures 3.35a and b). Calculation of $\Delta\delta^{RS}$ (Figure 3.35c) gives a negative difference for the methyl substituent and a positive difference for the isobutyl group. Comparison with the graphical model of the correlation locates the substituent with negative $\Delta\delta^{RS}$ [i.e., Me(1′)] in the $L_1$ location (Figure 3.35d), while the substituent with positive $\Delta\delta^{RS}$ (i.e., the isobutyl group) is in the $L_2$ location. This leads to the configuration shown in Figure 3.35d, the (S) isomer.

Dreiding models representing the main conformers of the (R)- and the (S)-MPA derivatives (Figure 3.36) illustrate the positions of the $L_1/L_2$ groups relative to the phenyl group, and they explain the shieldings observed in each case and the meaning of the positive and negative $\Delta\delta^{RS}$ signs.

Note that in ketone cyanohydrins [48], unlike with aldehyde cyanohydrins [47], the CN carbon lies on the bisecting plane, and therefore the aromatic shielding

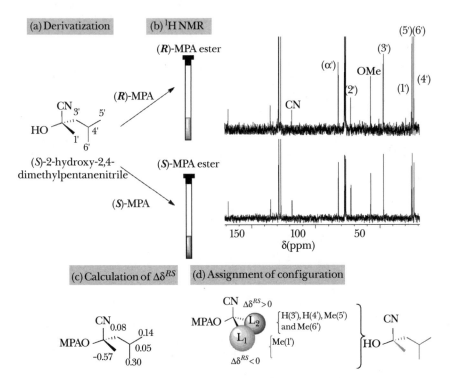

*Figure 3.35.* Main steps in the configurational assignment of (S)-2-hydroxy-2, 4-dimethylpentanenitrile using MPA and $^{13}$C NMR.

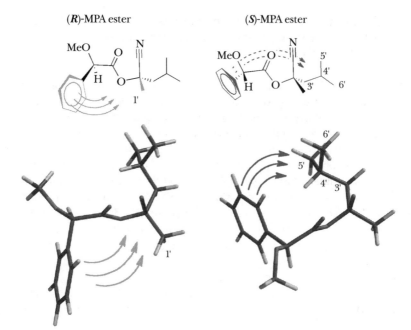

*Figure 3.36.* Position of the diagnostic carbons with regard to the phenyl group in the *sp* conformers of the two MPA esters of (S)-2-hydroxy-2,4-dimethylpentanenitrile.

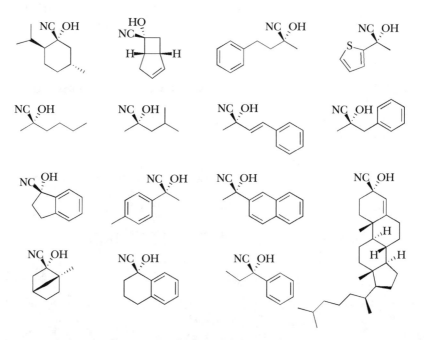

*Figure 3.37.* Selection of ketone cyanohydrins with known absolute configuration used to validate the procedures described in this chapter.

effect on this carbon is very similar in both derivatives. For that reason, this signal cannot be used for assignment [48].

### 3.4.4. Summary

Similar to the case with aldehyde cyanohydrins [47], the absolute configuration of ketone cyanohydrins [48, 49] can be assigned by double derivatization with (R)- and (S)-MPA. In this case, the main conformation determines that the CN group lies on the bisecting plane of the main conformations, and therefore is not useful for assignment. The signals for diagnosis are now the $L_1$ and $L_2$ substituents that occupy both sides of the plane in the same way as in secondary alcohols. Therefore, the assignment can be carried out using the $^1H$ $\Delta\delta^{RS}$ signs and/or the $^{13}C$ $\Delta\delta^{RS}$ signs of $L_1$ and $L_2$. The correlation between the spatial position of $L_1/L_2$ and their $\Delta\delta^{RS}$ sign is the same for both protons and carbons.

Figure 3.37 shows a selection of ketone cyanohydrins that have been used to validate this procedure [48, 49].

## ■ 3.5. SECONDARY THIOLS

The assignment of the absolute configuration of secondary thiols can be carried out following the same trends as were found for secondary alcohols, by derivatization with adequate carboxylic acids as the CDAs, followed by comparison of the NMR spectra of the corresponding (R)- and (S)-thioester derivatives [50–51].

Two CDAs have proven to be effective for this purpose: the most used and reliable ones are MPA and 2-naphtyl-*tert*-butoxyacetic acid (2-NTBA) [50, 51]. The first one is commercially available; and although the second one must be synthesized, it has the advantage of producing much larger shift differences.

Figure 3.38 shows their structures, the basic experimental conditions, the correlations between the $\Delta\delta^{RS}$ signs and the spatial location of $L_1/L_2$. This location is the same for both CDAs.

### 3.5.1. MPA and 2-NTBA Thioesters of Secondary Thiols

The replacement of the oxygen of a MPA ester by sulfur, which takes place in the corresponding MPA thioester, introduces relevant changes in the conformational equilibrium, which is now dominated by an *ap* conformer [50, 51], since the *sp* one is less stable. In the rapid conformational equilibrium, the *ap* form is the dominant one, and therefore the NMR spectra can be interpreted as if the structure of the MPA thioester were represented by that conformer. Figure 3.39 represents the *ap* conformations of the (R)- and (S)-MPA thioesters of a chiral thiol.

In this way, the phenyl group of MPA selectively shields the $L_2$ in the (R)-MPA thioester (Figure 3.39a) and the $L_1$ in the (S)-MPA thioester [50, 51] (Figure 3.39b).

Replacement of the phenyl and methoxy groups by 2-naphtyl and *tert*-butoxy groups, respectively, leads to 2-NTBA thioesters with basically the same *ap/sp* conformational composition as that of the MPA thioesters (Figures 3.39c and d);

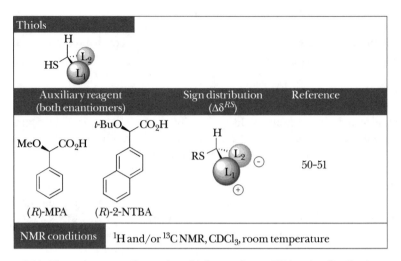

*Figure 3.38.* The assignment of secondary thiols at a glance: CDAs, sign distribution ($\Delta\delta^{RS}$), experimental conditions, and references.

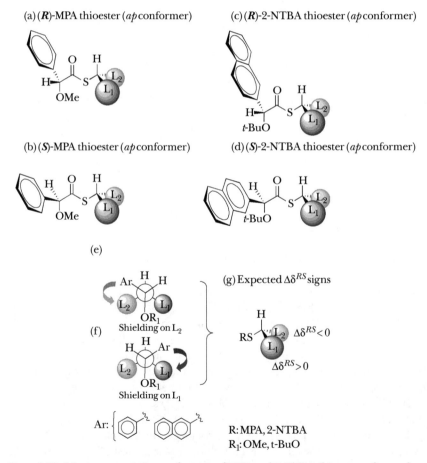

*Figure 3.39.* Most representative conformers of MPA and 2-NTBA thioesters of secondary thiols. Shielding effects and expected $\Delta\delta^{RS}$ signs are also shown.

there is also a higher predominance of the *ap* conformer over the *sp* one. This fact, together with the stronger shielding due to the naphthyl group, leads to larger $\Delta\delta^{RS}$ differences than those caused by MPA [50, 51].

The selective shieldings observed for $L_1/L_2$ in MPA and 2-NTBA thioesters (Figures 3.39e and f) lead to the same $\Delta\delta^{RS}$ sign distribution for the two CDAs; this can be seen in the stereomodel shown in Figure 3.39g: positive $\Delta\delta^{RS}$ for $L_1$ and negative $\Delta\delta^{RS}$ for $L_2$.

Both the $^1$H- and $^{13}$C-NMR chemical shifts of MPA or 2-NTBA derivatives of a thiol can be used for the assignment.

Examples illustrating applications of this procedure follow in Sections 3.5.2 to 3.5.4.

### 3.5.2. Example 13: Assignment of the Absolute Configuration of (*S*)-Butane-2-Thiol Using MPA

A sample of a pure enantiomer of butane-2-thiol, known by independent means to be the (*S*)-enantiomer, is separately derivatized with (*R*)- and (*S*)-MPA, and the $^1$H-NMR spectra of the MPA thioesters are registered and compared (Figures 3.40a and b). The signals are assigned by paying special attention to the protons of the thiol part located on either side of the *ap* plane. Thus, the signals for diagnosis are Me(1') on one side and the H(3')/Me(4') protons on the other (Figure 3.40b).

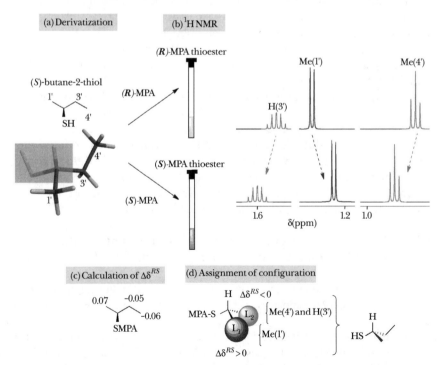

*Figure 3.40.* Main steps in the configurational assignment of (*S*)-butane-2-thiol using MPA.

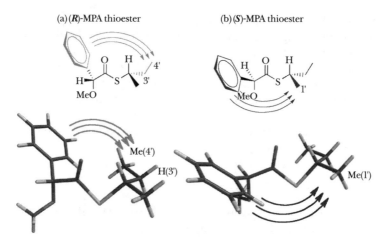

*Figure 3.41.* Position of the diagnostic hydrogens with regard to the phenyl group in the *ap* conformers of the two MPA thioesters of (S)-butane-2-thiol.

As shown in Figure 3.40, Me(1′) is more shielded in the (S)- than in the (R)-MPA derivative, while H(3′)/Me(4′) are more shielded in the (R)- than in the (S)-MPA derivative.

Subtraction of the chemical shifts leads to the $\Delta\delta^{RS}$ signs shown in Figure 3.40c: negative for H(3′)/Me(4′) and positive for Me(1′). Application of the sign/spatial location correlation for MPA thioesters of secondary thiols (Figure 3.40d) places the H(3′)/Me(4′) substituent behind the tetrahedron (i.e., $L_2$) and Me(1′) at the front (i.e., $L_1$). This leads to the (S) stereochemistry, coincident with that of the initial sample.

A full explanation of the spectra and a rationalization of the observed shifts and $\Delta\delta^{RS}$ signs can be obtained by inspection of Dreiding models of the (R)- and (S)-MPA derivatives representing the *ap* conformers, as shown in Figure 3.41.

As can be observed in Figure 3.41a, H(3′)/Me(4′) are shielded in the *ap* conformer of the (R)-MPA derivative and are not affected in the (S)-MPA derivative. On the other hand, Me(1′) is shielded in the *ap* conformer of the (S)-MPA derivative and is not affected in the (R)-MPA derivative. This leads to the observed negative value for $\Delta\delta^{RS}$ for H(3′)/Me(4′) and the positive $\Delta\delta^{RS}$ for Me(1′).

### 3.5.3. Example 14: Assignment of the Absolute Configuration of (R)-Ethyl 2-Mercaptopropanoate Using 2-NTBA

Similar to what was done in the preceding section, the absolute configuration of (R)-ethyl 2-mercaptopropanoate was checked by NMR using 2-NTBA as the CDA (Figure 3.42).

Drawing of the plane defined by the *ap* conformer allows us to identify the diagnostic protons, which in this case are Me(3′) on one side and H(4′)/Me(5′) on the other (Figures 3.42a and b). Comparison of their chemical shifts indicates that H(4′)/Me(5′) are more shielded in the (S)- than in the (R)-2-NTBA derivative,

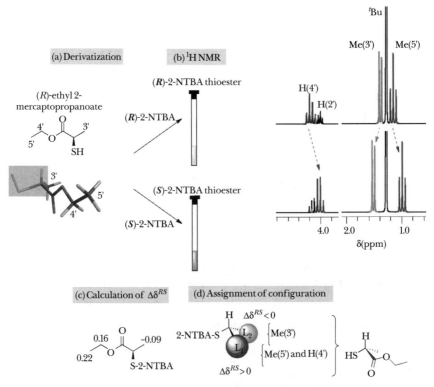

*Figure 3.42.* Main steps in the configurational assignment of (R)-ethyl 2-mercaptopropanoate using 2-NTBA.

while Me(3′) is more shielded in the (R)- than in the (S)-2-NTBA derivative. These spectra lead to the $\Delta\delta^{RS}$ differences and signs shown in Figure 3.42c. When these are applied to the sign/spatial location correlation model, they lead to the stereochemistry shown in Figure 3.42d, which is coincident with that of the original sample. As previously indicated, 2-NTBA produces the same sign distribution as does MPA, but the $\Delta\delta^{RS}$ values are clearly larger. Therefore 2-NTBA is particularly appropriate when MPA yields very small differences, and no sure sign can be derived from those values.

Inspection of Dreiding models of the (R)- and the (S)-2-NTBA thioesters representing the main *ap* conformation (Figure 3.43) allows the immediate rationalization of those shifts just by analyses of the relative location of $L_1/L_2$ with regard to the naphthyl group responsible for the shieldings.

### 3.5.4. Example 15: Assignment of the Absolute Configuration of (R)-Ethyl 2-Mercaptopropanoate Using 2-NTBA and $^{13}$C NMR

A sample of ethyl 2-mercaptopropanoate of unknown configuration is separately derivatized with (R)- and (S)-2-NTBA, and the $^{13}$C-NMR spectra of the

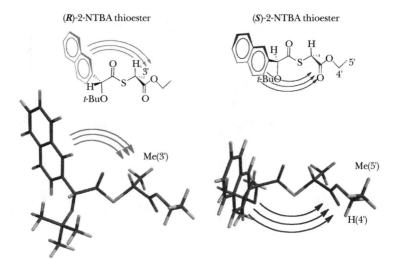

*Figure 3.43.* Position of the diagnostic hydrogens with regard to the naphthyl group in the *ap* conformers of the two 2-NTBA thioesters of (R)-ethyl 2-mercaptopropanoate.

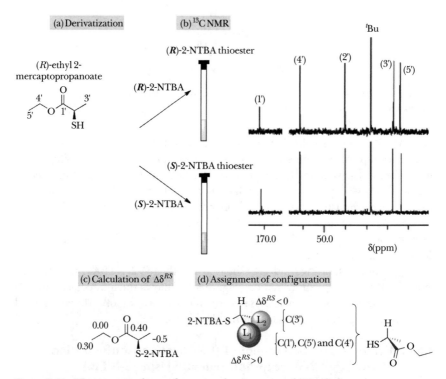

*Figure 3.44.* Main steps in the configurational assignment of (R)-ethyl 2-mercaptopropanoate using 2-NTBA and $^{13}$C NMR.

thioesters is recorded and assigned [50, 51, 72] (Figures 3.44a and b). Inspection of the structure indicates that the carbon signals useful for diagnosis are the methyl [C(3′)] group on one side and the carbonyl [C(1′)] and ethoxy [C(4′)/C(5′)] groups on the other. Visual inspection of the spectra indicates that some carbons are slightly more shielded in one derivative than in the other. The differences in $\Delta\delta^{RS}$ (Figure 3.44c) are negative for C(3′) and positive for C(1′)/C(4′)/C(5′). Thus, in accordance with the sign/spatial location correlation for secondary thiols (Figure 3.44d), the methyl substituent [C(3′)] must be placed at the back (negative $\Delta\delta^{RS}$, i.e., $L_2$) and the ethoxy-carbonyl substituent [C(1′)/C(4′)/C(5′)] at the front of the tetrahedral model (positive $\Delta\delta^{RS}$, i.e., $L_1$). This stereochemical disposition corresponds to the (R)-enantiomer.

Figure 3.43 represents the main conformers (*ap*) of the (R)- and (S)-2-NTBA thioester derivatives and the position of the methyl and ethoxycarbonyl groups: one on each side of the bisecting plane and therefore with different positions relative to the naphthyl ring of the auxiliary. The methyl [C(3′)] is under the shielding cone in the (R)-2-NTBA derivative, while the ethoxycarbonyl substituent [C(1′)/C(4′)/C(5′)] is shielded in the (S)-2-NTBA derivative. This situation leads to $\Delta\delta^{RS}$ signs coincident with the experimental ones found for the (R)-enantiomer.

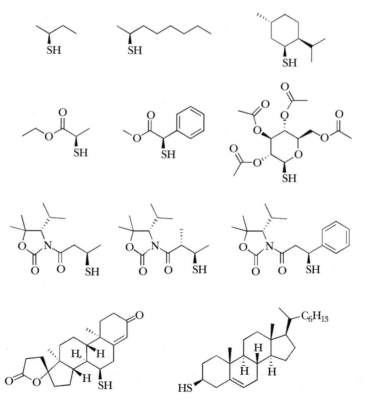

*Figure 3.45.* Selection of secondary thiols with known absolute configuration used to validate the procedures described in this chapter.

## 3.5.5. Summary

The absolute configuration of secondary thiols can be determined by double derivatization using either MPA or 2-NTBA and followed by comparison of the NMR spectra of the (R)- and the (S)-thioester derivatives [50, 51]. MPA is commercially available and therefore easier to use, but in general, it yields small $\Delta\delta^{RS}$ values, which in some cases are not large enough to provide significant positive or negative signs. Replacements of methoxy by *tert*-butoxy and phenyl by 2-naphtyl groups (i.e., MPA by 2-NTBA) provide the corresponding thioester derivatives with a more favored conformation and stronger aromatic shieldings, thus allowing a safe assignment in all the cases tested. The main conformation with both CDAs is *ap*, and the signals for diagnosis are those from the $L_1/L_2$ substituents.

The assignment of thiols can be carried out from either/both $^1$H- or/and $^{13}$C-NMR spectra, using the same correlation model to associate the spatial position of $L_1/L_2$ with their $\Delta\delta^{RS}$ signs.

Figure 3.45 shows a selection of secondary thiols that have been used to validate this procedure [50, 51]. Examples of applications of this methodology can be found in the literature [148].

## 3.6. α-CHIRAL PRIMARY AMINES

The assignment of α-chiral primary amines by double derivatization can be carried out making use of several CDAs [14, 15, 52–54]. The most reliable ones are BPG [54], MPA [52] and MTPA [53]. Figure 3.46 shows their structure, the corresponding correlations between their signs and spatial locations, and gives a summary of the experimental conditions. The preparation of the derivatives is nearly identical for the three reagents. BPG is cheaper and generally results in larger $\Delta\delta^{RS}$ values than those for the other two auxiliaries [54], but any of the three can be safely used.

### 3.6.1. BPG as the CDA for α-Chiral Primary Amines

The conformational composition of BPG amides [54] consist of *ap* and *sp* forms in equilibrium, the *ap* one being the most stable. The NMR spectrum of the equilibrium can be easily interpreted by assuming that the *ap* conformation is the representative structure (Figure 3.47). The *ap* conformer has the Cα–H, C=O and Cα'-H in a coplanar arrangement, with the Cα–H and C=O bonds *anti* and the C=O and Cα'-H *syn* [54] (Figures 3.47a and b).

For this reason, the shielding due to the phenyl group affects only the $L_1/L_2$ substituent located on the same side of the plane. Thus, $L_1$ is more shielded in the (R)- than in the (S)-BPG amide, while $L_2$ is more shielded in the (S)- than in the (R)-BPG amide [54]. Therefore, a negative $\Delta\delta^{RS}$ will be expected for $L_1$ and a positive $\Delta\delta^{RS}$ for $L_2$ (Figure 3.47c).

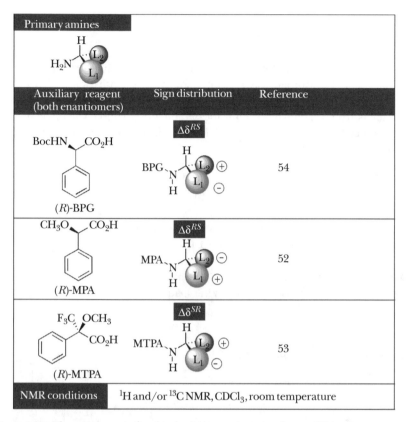

*Figure 3.46.* The assignment of α-chiral primary amines at a glance: CDAs, sign distributions ($\Delta\delta^{RS}$, $\Delta\delta^{SR}$), experimental conditions, and references.

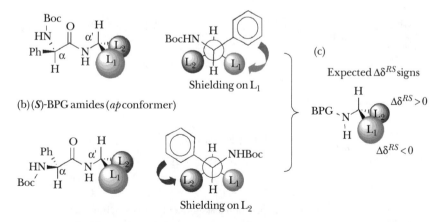

*Figure 3.47.* Shielding effects in the most representative conformers of BPG amides of α-chiral primary amines and expected $\Delta\delta^{RS}$ signs.

## 3.6.2. Example 16: Assignment of the Absolute Configuration of (−)-Isopinocampheylamine Using BPG

A sample of a pure enantiomer of isopinocampheylamine, known to be the (1$R$,2$R$,3$R$,5$S$)-(−)-isomer, was separately derivatized with ($R$)- and with ($S$)-BPG. Figure 3.48b shows the partial $^1$H-NMR spectra of the amides containing signals from diagnostic protons located on either side of the plane defined by the *ap* conformation: H(2′)/H(3′)/Me(8′) on one side and H(5′)/H(6′) on the other.

Comparing the ($R$)- to the ($S$)-BPG derivatives (Figure 3.48b), we observe that the H(2′)/H(3′)/Me(8′) signals move to a lower field (negative $\Delta\delta^{RS}$), while the H(5′)/H(6′) move to a higher field (positive $\Delta\delta^{RS}$). Application of the correlation model (Figure 3.47c) places H(2′)/H(3′)/Me(8′) at the $L_1$ position at the front of the tetrahedron around C$\alpha$′[i.e., C(1′)] and H(5′)/H(6′) at the rear ($L_2$), leading to the stereochemistry shown in Figure 3.48d.

Dreiding models representing the *ap* conformation of BPG amides allow (Figure 3.49) a quick rationalization of the shifts observed in the spectra and the correlation between the $\Delta\delta^{RS}$ signs and the spatial location.

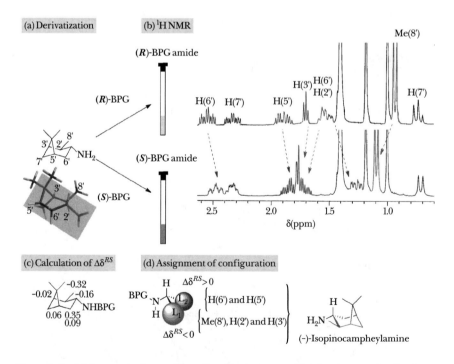

*Figure 3.48.* Main steps in the configurational assignment of (−)-isopinocampheylamine using BPG.

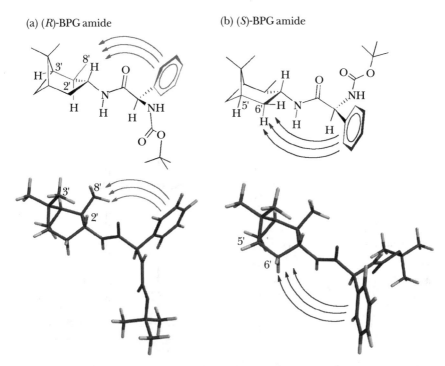

*Figure 3.49.* Position of the diagnostic hydrogens with regard to the phenyl group in the *ap* conformers of the two BPG amides of (−)-isopinocampheylamine.

### 3.6.3. Example 17: Assignment of the Absolute Configuration of (*S*)-Butan-2-Amine Using BPG and $^{13}$C NMR

$^{13}$C NMR chemical shifts of BPG amides can also be used for the assignment of chiral amines [72] by treating the $^{13}$C $\Delta\delta^{RS}$ signs in exactly the same way as was done for the $^1$H $\Delta\delta^{RS}$ signs from the $^1$H-NMR spectra [54]. As an example to illustrate this, we show the $^{13}$C-NMR spectra of the BPG amide of (*S*)-butan-2-amine (Figure 3.50). $^{13}$C-NMR data indicate that the carbons C(3′)/C(4′) present positive $\Delta\delta^{RS}$ signs, while the carbon C(1′) presents a negative one (Figure 3.50c). The model shown in Figures 3.47c and 3.50d places the C(3′)/C(4′) in the $L_2$ position and the carbon C(1′) in the $L_1$ position. Thus, the absolute stereochemistry of the amine corresponds to (*S*)-butan-2-amine (Figure 3.50d).

If we build Dreiding models representing the main conformers of the (*R*)- and (*S*)-BPG amide derivatives (Figure 3.51), we can easily observe the different positions of the two substituents of the asymmetric carbon, methyl and ethyl groups [C(1′) and C(3′)/C(4′), respectively], relative to the phenyl group of the auxiliary: C(1′) is under the shielding cone in the (*R*)-BPG derivative but not so in the (*S*)-BPG derivative, and the opposite situation happens with C(3′)/C(4′). This explains the shieldings observed in each case and the negative [i.e., C(1′)] and positive [i.e., C(3′)/C(4′)] $\Delta\delta^{RS}$ signs.

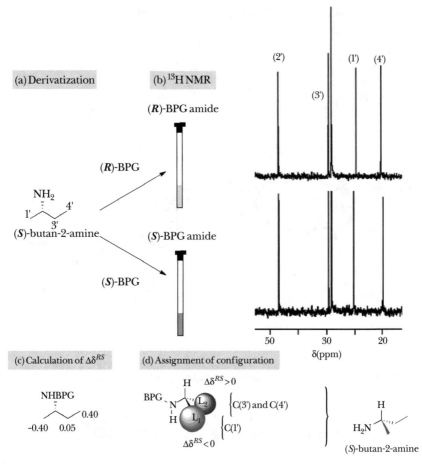

*Figure 3.50.* Main steps in the configurational assignment of (S)-butan-2-amine using BPG and $^{13}$C NMR.

### 3.6.4. MPA as the CDA for α-Chiral Primary Amines

In a similar way, the absolute configuration of an amine can be obtained by comparison of the (R)- and (S)-MPA amides [52]. With this CDA, we should keep in mind that *ap* is again the representative conformer: the Cα–OMe, C=O and Cα'-H bonds are coplanar, with the Cα–OMe/C=O and C=O/Cα'-H bonds in *anti* and *syn* arrangements, respectively [52] (Figure 3.52). This leads to a sign distribution [52] in which the L substituent with a positive $\Delta\delta^{RS}$ is at the front of the tetrahedron around Cα' (i.e., $L_1$; Figure 3.52c), and the L substituent with a negative $\Delta\delta^{RS}$ is in the back (i.e., $L_2$; Figure 3.52c).

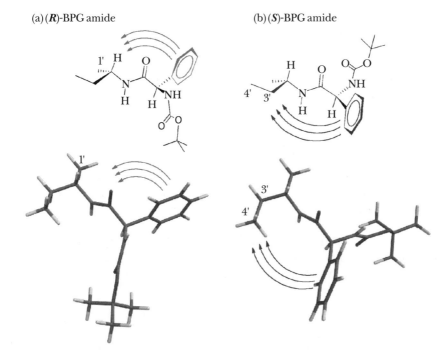

*Figure 3.51.* Position of the diagnostic carbons with regard to the phenyl group in the *ap* conformers of the two BPG amides of (*S*)-butan-2-amine.

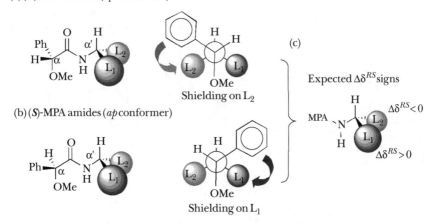

*Figure 3.52.* Shielding effects in the most representative conformers of MPA amides of α-chiral primary amines and expected $\Delta\delta^{RS}$ signs.

### 3.6.5. Example 18: Assignment of the Absolute Configuration of (−)-Bornylamine Using MPA

A sample of a pure enantiomer of bornylamine, known to be the (−)-isomer, was separately derivatized with (R)- and with (S)-MPA. Figure 3.53 shows the partial $^1$H-NMR spectra of the derivatives containing signals for diagnostic protons located on either side of the plane defined by the *ap* conformation: H(3′)/Me(10′) on one side and H(5′)/H(6′) on the other.

When comparing the (R)- with the (S)-MPA derivative, the signals from H(3′)/Me(10′) resonate at a lower field (negative $\Delta\delta^{RS}$), while those from H(5′)/H(6′) resonate at a higher field (positive $\Delta\delta^{RS}$). Application of the correlation model (Figure 3.52c) places the H(3′)/Me(10′) at the back of the tetrahedron (i.e., $L_2$) and H(5′)/H(6′) in the front (i.e., $L_1$), leading to the stereochemistry shown in Figure 3.53d.

Rationalization of the chemical shifts observed for each derivative and explanations of the resulting correlation between the $\Delta\delta^{RS}$ signs and the spatial locations are obtained by examination of Dreiding models representing the *ap* conformations (Figure 3.54).

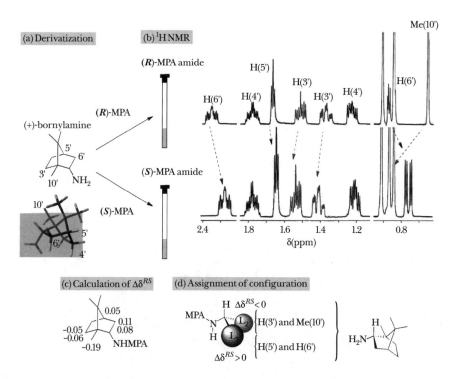

*Figure 3.53.* Main steps in the configurational assignment of (−)-bornylamine using MPA.

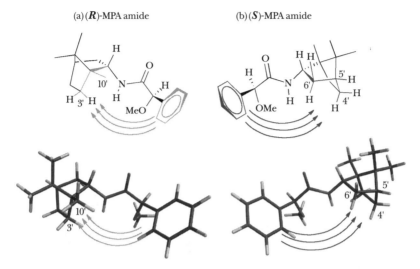

*Figure 3.54.* Position of the diagnostic hydrogens with regard to the phenyl group in the *ap* conformers of the two MPA amides of (−)-bornylamine.

### 3.6.6. MTPA as the CDA for α-Chiral Primary Amines

Amines can also be derivatized as MTPA amides, and their absolute configuration can be assigned by comparison of the derivatives [53]. As in the case of MTPA esters, MTPA amides are also composed of three main conformers: two of the *ap* and one of the *sp* types [53]. From the NMR point of view, the *sp* conformation is the most representative, and it can be used to explain and predict the chemical shifts and the shieldings of $L_1/L_2$. Figure 3.55 shows the conformations, the corresponding shieldings (Figures 3.55a and b), and the resulting correlation between the $\Delta\delta^{SR}$ signs and the spatial locations [53] (Figure 3.55c), which places the L substituent with negative $\Delta\delta^{SR}$ in the front of the tetrahedron (i.e., $L_1$; Figure 3.55c) and the substituent with positive $\Delta\delta^{SR}$ at the back (i.e., $L_2$; Figure 3.55c). It is necessary to remember that in the MTPA derivatives we use $\Delta\delta^{SR}$ instead of $\Delta\delta^{RS}$ [38, 53].

### 3.6.7. Example 19: Assignment of the Absolute Configuration of (−)-Bornylamine Using MTPA

A sample of a pure enantiomer of bornylamine, known to be the (−)-isomer, was separately derivatized with (R) and (S)-MTPA. The Figure 3.56b shows the partial spectra of the derivatives containing signals from the diagnostic protons located on either side of the plane defined by the *sp* conformation: H(3′)/Me(10′) on one side and H(5′)/H(6′) on the other.

When we go from the spectrum of the (S)- to that of the (R)-MTPA derivative, the signals from H(3′)/Me(10′) resonate at a higher field (positive $\Delta\delta^{SR}$), while those of H(5′)/H(6′) resonate at a lower field (negative $\Delta\delta^{SR}$). Application of the

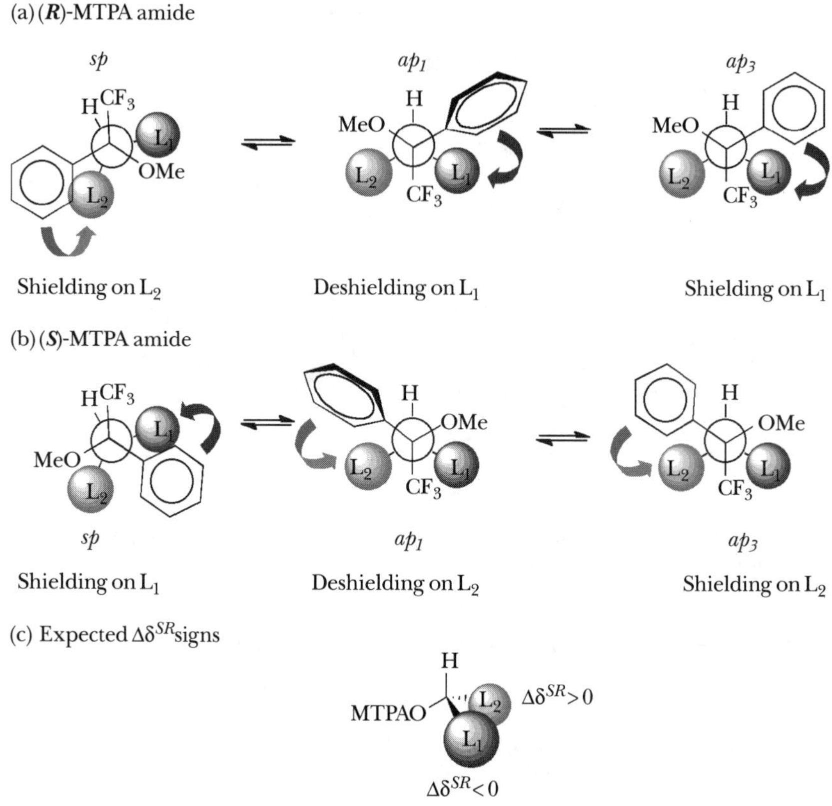

(a) (*R*)-MTPA amide

*Figure 3.55.* Main conformations of the MTPA amides of α-chiral primary amines. Shielding/deshielding effects in the most representative conformers and expected $\Delta\delta^{SR}$ signs.

correlation model (Figure 3.55c) places H(3′)/Me(10′) behind the tetrahedron (i.e., $L_2$ in Figure 3.56d) and H(5′)/H(6′) at the front (i.e., $L_1$ in Figure 3.56d), leading to the stereochemistry shown in Figure 3.56d.

Rationalization of the chemical shifts observed for each derivative and explanation of the resulting correlation between the $\Delta\delta^{SR}$ signs and the spatial locations are obtained by examination of Dreiding models representing the *sp* conformation (Figure 3.57).

## 3.6.8. Summary

As we have shown in the preceding pages, the absolute configuration of chiral monosubstituted amines can be determined by double derivatization with one of three reagents (MPA, MTPA, and BPG), followed by comparison of the chemical shifts for $L_1$ and $L_2$ in the (R)- and the (S)-amide derivatives [14, 15, 52–54]. The three CDAs are commercially available and provide $\Delta\delta^{RS}/\Delta\delta^{SR}$ values that, in most

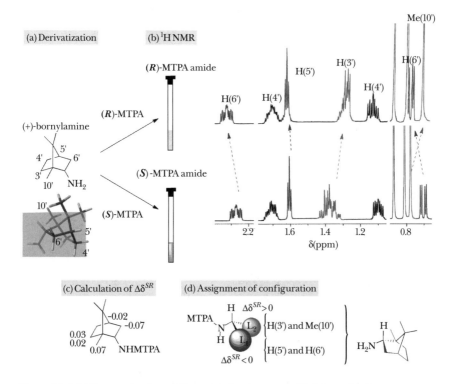

*Figure 3.56.* Main steps in the configurational assignment of (−)-bornylamine using MTPA.

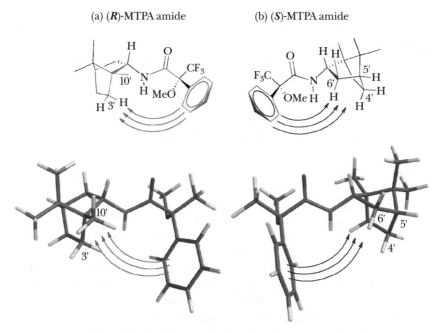

*Figure 3.57.* Position of the diagnostic hydrogens with regard to the phenyl group in the *ap* conformers of the two MPA amides of (−)-bornylamine.

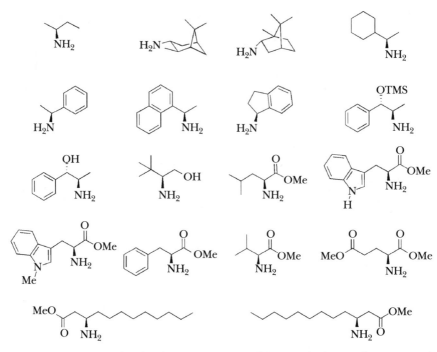

*Figure 3.58.* Selection of α-chiral primary amines with known absolute configuration used to validate the procedures described in this chapter.

cases, are large enough to produce significant values and homogeneously distributed signs, thus allowing a safe assignment.

Although the amides derived from these three reagents share similar types of *ap/sp* conformational equilibria, the main conformers (and therefore the correlation between the spatial locations of $L_1/L_2$ and the chemical shifts) are different (Figure 3.46).

It is important to mention that for historical reasons, when MTPA is used, the difference between the chemical shifts is measured as $\Delta\delta^{SR}$ instead of $\Delta\delta^{RS}$, as with all of the other CDAs [38, 53].

In order to prevent confusion [85], it should be stressed that when preparing MTPA amides, the acid and the acid chloride result in compounds with different nomenclature; this is due to the Cahn-Ingold-Prelog (CIP) priority rules, which give different priorities to the substituents of MTPA and MTPA chloride. Thus, the (R)-enantiomer of MTPA-Cl produces the (S)-MTPA amide, while (R)-MTPA produces the (R)-MTPA amide [85], and the reverse is true for the (S)-enantiomers.

From a practical point of view, BPG has some advantages [54] over the other two compounds, namely, it is cheaper and produces $\Delta\delta^{RS}$ values that, in general, are larger than those obtained with MPA [52] or MTPA [53].

These amines can be assigned by $^{13}$C NMR of the very same derivatives utilized for $^1$H NMR and under the same correlations between the $\Delta\delta^{RS}$ values and the spatial locations of $L_1$ and $L_2$ [72]. In this way, recording the $^1$H- and $^{13}$C-NMR spectra of the samples provides ¹H and ¹³C $\Delta\delta^{RS}$ signs that can be used either together or separately for the assignment. The ¹³C $\Delta\delta^{RS}$ values are very small compared to those obtained from ¹H NMR, but in most cases, the signs are sufficiently large that an assignment can be made [72].

Figure 3.58 shows a selection of the amines that have been used to validate the procedures described in this chapter [52–54]. Other examples of applications of this methodology can be found in the literature [149–160].

Configurational assignment using only one derivative [55–56], either the (R)- or the (S)-MPA amide, is also possible, as will be described in Chapter 4.

## ■ 3.7. α-CHIRAL CARBOXYLIC ACIDS

The assignment of the absolute configuration of α-chiral carboxylic acids can be carried out in a similar way [14, 15, 57, 58] to that described for other chiral compounds (e.g., secondary alcohols), but obviously, since the substrate now has a carboxyl group available for bonding (i.e., instead of a hydroxyl group), a different auxiliary reagent is required [57, 58]. The CDA of choice for carboxylic acids is 9-anthrylhydroxyacetic acid ethyl ester (9-AHA), which has characteristics that are similar to 9-AMA, but with the carboxyl group blocked as ethyl ester and a free hydroxyl functional group that is used for bonding to the substrates.

The assignment requires the preparation of the corresponding (R)- and (S)-9-AHA esters, and the $\Delta\delta^{RS}$ signs for $L_1/L_2$ are used to determine the locations of those substituents [57, 58].

The ¹³C-NMR chemical shifts of the 9-AHA esters can also be used for the assignment of the chiral acid [72] by treating the ¹³C $\Delta\delta^{RS}$ signs in exactly the same way as was done for the proton $\Delta\delta^{RS}$ signs.

Figure 3.59 presents a summary of this methodology.

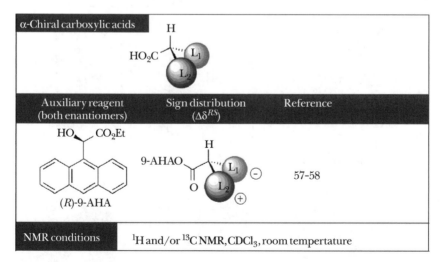

*Figure 3.59.* The assignment of α-chiral carboxylic acids at a glance: CDA, sign distribution ($\Delta\delta^{RS}$), experimental conditions, and references.

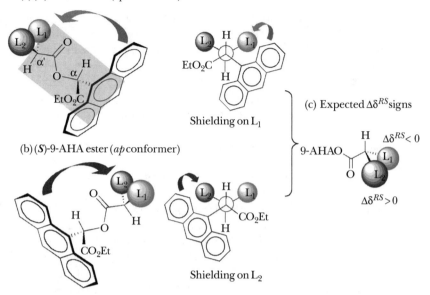

*Figure 3.60.* Shielding effects in the most representative conformers of 9-AHA esters of α-chiral carboxylic acids and expected $\Delta\delta^{RS}$ signs.

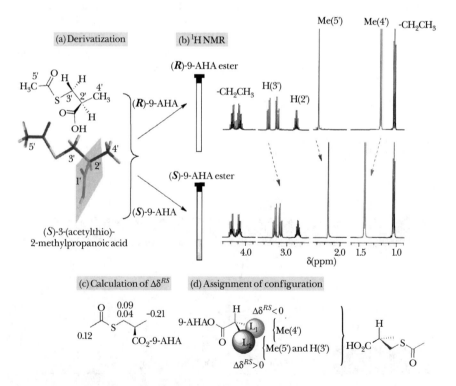

*Figure 3.61.* Main steps in the configurational assignment of (*S*)-3-(acetylthio)-2-methylpropanoic acid using 9-AHA.

## 3.7.1. 9-AHA Esters of Carboxylic Acids

From a conformational point of view, 9-AHA esters of carboxylic acids are composed of *sp* and *ap* conformers [57, 58], the latter being the most stable ones. Therefore, placing 9-AHA derivatives in the *ap* conformation allows a direct interpretation of the spectra and, in particular, of the shifts for $L_1$ and $L_2$.

It is necessary to point out that in these esters, rotation around the $(O_2)C-C\alpha'$ bond of the carboxylic acid moiety defines the *sp* and *ap* conformations: in *sp*, $C\alpha'$-H is synperiplanar to C=O, and in *ap* it is antiperiplanar [57, 58].

Figure 3.60 shows that when the (*R*)-9-AHA derivative is in the *ap* conformation, $L_1$ is under the shielding cone of the anthryl group. In the (*S*)-AHA derivative, $L_2$ is the most shielded substituent. Those differences produce the expected shifts and positive/negative sign distributions (negative $\Delta\delta^{RS}$ for $L_1$ and positive for $L_2$) that correlate with the spatial positions of $L_1/L_2$, as shown in Figure 3.60c.

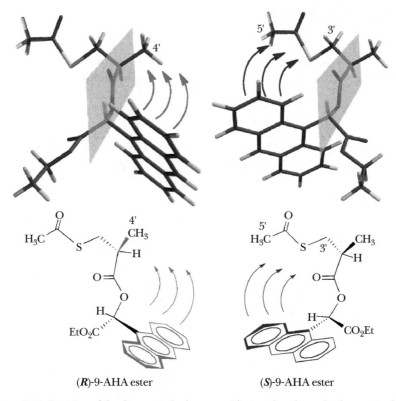

(*R*)-9-AHA ester       (*S*)-9-AHA ester

*Figure 3.62.* Position of the diagnostic hydrogens with regard to the anthryl group in the *ap* conformers of the two 9-AHA esters of (*S*)-3-(acetylthio)-2-methylpropanoic acid.

## 3.7.2. Example 20: Assignment of the Absolute Configuration of (S)-3-(Acetylthio)-2-Methylpropanoic Acid Using 9-AHA

A sample of a pure enantiomer of the chiral methylpropanoic acid, indicated in Figure 3.61 and known by independent means to be the (S)-enantiomer, was separately derivatized with (R)- and (S)-9-AHA, producing the corresponding 9-AHA esters (Figure 3.61a). Figure 3.61b shows the partial ¹H-NMR spectra containing signals from the diagnostic protons located on either side of the plane defined by the *ap* conformation, with Me(4′) on one side and H(3′)/Me(5′) on the other.

When we go from the spectrum of the (R)- to that of the (S)-9-AHA derivative, we observe that the signal for Me(4′) moves to a lower field (negative $\Delta\delta^{RS}$; Figure 3.61c) while the H(3′)/Me(5′) protons move to a higher field (positive $\Delta\delta^{RS}$; Figure 3.61c). Application of the correlation model (Figure 3.61d) places Me(4′) at the back of the tetrahedron around C(2′) in the $L_1$ position, and the H(3′)/Me(5′) protons at the front in the $L_2$ position. This leads to the stereochemistry shown in Figure 3.61d, which corresponds to the (S)-enantiomer.

Dreiding models of the (R)- and the (S)-9-AHA ester derivatives in the *ap* conformation (Figure 3.62) clearly show the plane bisecting the structure, the spatial location of the substituents, and the selectivity of the aromatic shieldings. They explain the spectra and the configuration assigned to this substrate.

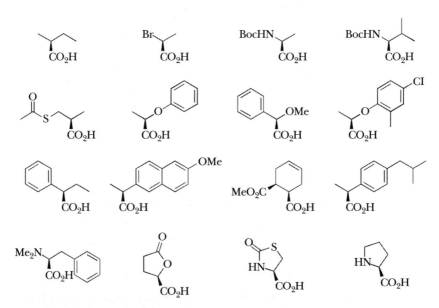

*Figure 3.63.* Selection of α-chiral carboxylic acids with known absolute configuration used to validate the procedures described in this chapter.

### 3.7.3. Summary

The absolute configuration of α-chiral carboxylic acids can be determined by double derivatization with (R)- and (S)-9-AHA and comparison of the resulting NMR spectra [57, 58]. The main conformation of (R)- and (S)-9-AHA esters is *ap*, and this places the $L_1$ and $L_2$ groups on either side of the bisecting plane. This geometry renders the signals due to $L_1$ and $L_2$ sensitive to the aromatic shielding of the reagent, and thus it is valid for stereochemical diagnosis. Because the correlation between the spatial positions of $L_1/L_2$ and their $\Delta\delta^{RS}$ signs is common for the $^1$H and $^{13}$C chemical shifts, both types of nuclei can be used for the assignment [57, 58, 72].

Figure 3.63 shows a selection of carboxylic acids with known absolute configurations that have been used to validate this procedure [57, 58]. Other examples of applications can be found in the literature [161–164].

# 4 Assignment of the Absolute Configuration of Monofunctional Compounds by Single Derivatization

The procedures shown in Chapter 3 allow the determination of the absolute configuration of several classes of compounds (Chapter 1, Figure 1.18), but they require the preparation of two derivatives and the comparison of their NMR spectra.

Alternative methods have been developed for secondary alcohols and α-chiral primary amines. These are particularly suited for those cases where the amount of the available sample is low, and they require the preparation of only a single derivative [41–43, 55–56, 165].

There are three different approaches to using only a single derivatization to perform the assignment of those substrates [13, 165]. The first two (Figures 4.1a and b) are based on a controlled conformational change that is produced either by modification of the probe temperature [41, 165] or by selective complexation [42, 55, 56, 165]. The third one (Figure 4.1c) is based on the differences observed between the chemical shifts of the free alcohol and those of the 9-AMA ester derivative [43, 165].

In general, these single-derivatization procedures are limited to $^1$H NMR. Because the shift differences observed in $^{13}$C NMR are quite small, they produce insignificant $\Delta\delta$ values, and therefore the signs are not sufficiently accurate to produce a safe assignment [72].

Explanations and examples of applications are presented in the remainder of this chapter.

## ■ 4.1. LOW-TEMPERATURE NMR PROCEDURE FOR SECONDARY ALCOHOLS

For the assignment of secondary alcohols, a simple approach based on the use of a single MPA ester has proven to work very well [41, 165]. It is based on the controlled shift of the conformational equilibrium between the two main conformers (*sp/ap*) that were described in Chapter 1 for the MPA esters of secondary alcohols [36, 37]. Thus, for the assignment, it is only necessary to prepare either the (*R*)- or the (*S*)-MPA ester and then to compare the chemical shifts of $L_1/L_2$ in the spectra taken at room temperature and at a lower temperature [41]. Figure 4.2 presents a summary of the procedure and the graphical model expressing the $\Delta\delta^{T1T2}$ correlation between the sign and the stereochemistry for the assignment of secondary alcohols derivatized as (*R*)- or as (*S*)-MPA esters.

(a) Lowering the NMR probe temperature

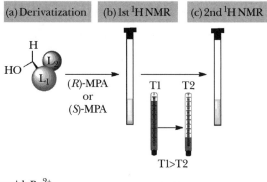

(b) Complexation with $Ba^{2+}$

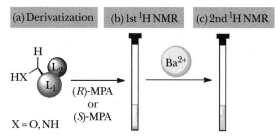

(c) Esterification shifts

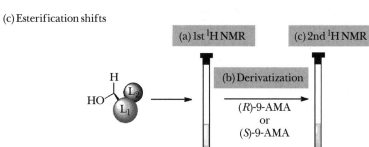

*Figure 4.1.* Methods to assign the absolute configuration of monofunctional compounds by single derivatization.

Figure 4.3 shows the conformational equilibrium [36, 37] (Figure 4.3a) and the idealized spectra of the (*R*)-MPA ester of a secondary alcohol registered at room temperature and at a lower temperature (Figures 4.3b–d). When the probe temperature is lowered, the conformational equilibrium is shifted in a way that increases the concentration of the most stable *sp* form and diminishes that of the least stable *ap* one [36, 41]. In this way, the number of molecules having the $L_1$ group under the shielding cone of the auxiliary increases at lower temperature, and the number of molecules having the $L_2$ group under the shielding cone of the auxiliary diminishes. As a result, the low-temperature ¹H-NMR spectra will show $L_1$ more shielded and $L_2$ less shielded than at room temperature (Figure 4.3d),

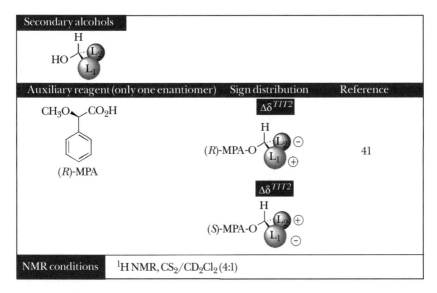

*Figure 4.2.* The assignment of secondary alcohols by the low-temperature NMR procedure at a glance: CDA, sign distributions ($\Delta\delta^{T1T2}$), experimental conditions, and references.

and the signs of the differences in the chemical shifts reflect the position of $L_1/L_2$ relative to the auxiliary [41]. This difference is defined as $\Delta\delta L^{T1T2} = \delta L(T_1) - \delta L(T_2)$, where $T_1 > T_2$. The signs of $\Delta\delta^{T1T2}$ for $L_1/L_2$ can be used to correlate the absolute configuration with the chemical shifts in the same way as was done with $\Delta\delta^{RS}$ (Figure 4.4).

Experimental data and theoretical calculations, as well as testing with a large series of secondary alcohols of known absolute configuration, have validated the correlation between the signs of $\Delta\delta^{T1T2}$ and the spatial location [41].

In general, temperatures lower than 200K are not required for this procedure. For the solvent, a 4:1 $CS_2/CD_2Cl_2$ mixture is recommended [41].

### 4.1.1. Example 21: Assignment of the Absolute Configuration of Diacetone D-Glucose Using (*R*)-MPA

We now consider application of this method [41] to the assignment of diacetone D-glucose, which is a representative example of a secondary alcohol (Figure 4.5).

Once the alcohol has been derivatized with (*R*)-MPA, the spectra of the ester at two different temperatures are recorded (Figure 4.5a). The partial $^1$H-NMR spectra of the (*R*)-MPA ester in $CD_2Cl_2/CS_2$ (1:4) at 300K and 193K are shown in Figures 4.5b and d, respectively. They include signals due to relevant protons at both sides of the asymmetric carbon: H(4')/H(5')/H(6') on one side and H(1')/H(2') on the other. Comparison between the spectra demonstrates that the signals due to H(4')/H(5')/H(6') shift to a higher field at the lower temperature (Figure 4.5d), and they have positive $\Delta\delta^{T1T2}$ values (Figure 4.5e). The signals

**96** ■ The Assignment of the Absolute Configuration by NMR

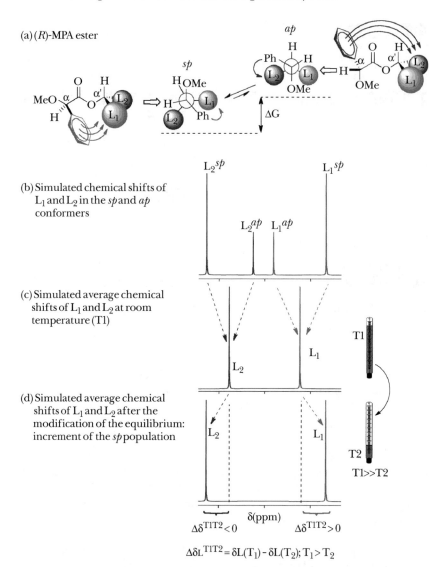

*Figure 4.3.* Conceptual representation of the low-temperature NMR procedure for secondary alcohols (single derivatization procedure) making use of an (*R*)-MPA ester.

due to H(1′)/H(2′) move to a lower field at the lower temperature, and they have negative $\Delta\delta^{T1T2}$ values (Figures 4.5d and e). Application of the graphical model for (*R*)-MPA esters [Figure 4.4a; the substituent with positive $\Delta\delta^{T1T2}$ is at the front ($L_1$), and the one with negative $\Delta\delta^{T1T2}$ is at the back ($L_2$)] places H(1′)/H(2′) at the back ($L_2$) and H(4′)/H(5′)/H(6′) at the front ($L_1$) of the tetrahedron, leading to the absolute configuration shown in Figure 4.5f.

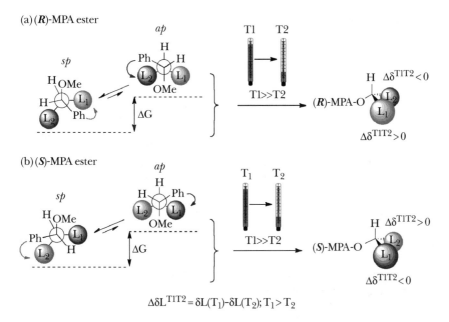

$\Delta\delta L^{T1T2} = \delta L(T_1) - \delta L(T_2); T_1 > T_2$

*Figure 4.4.* Conformational equilibria and $\Delta\delta^{T1T2}$ values for the MPA esters of secondary alcohols.

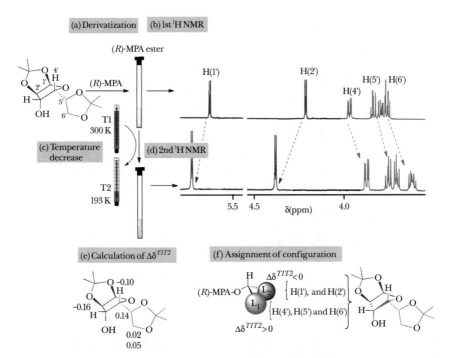

*Figure 4.5.* Main steps in the procedure for the configurational assignment of diacetone D-glucose, using (R)-MPA and low-temperature NMR.

### 4.1.2. Example 22: Assignment of the Absolute Configuration of (R)-Butan-2-ol Using (S)-MPA

Figure 4.6 shows a schematic of the procedure for the configurational assignment of an enantiomer of butan-2-ol, based on the $^1$H-NMR spectra of the (S)-MPA ester derivative. As before (see Section 4.1), from the assignment of the signals and comparison of the spectra registered at 300K and 203K, we see that Me(1′) resonates at a higher field at the lower temperature, while H(3′)/Me(4′) shift to lower field (Figure 4.6). Calculation of the differences of the chemical shifts ($\Delta\delta^{T1T2}$) gives the corresponding signs, and comparison with the model of Figure 4.4b allows us to place H(3′)/Me(4′) at $L_1$ and Me(1′) at $L_2$, leading to the (R) absolute configuration shown in Figure 4.6f.

Dreiding models explaining these trends observed in the $^1$H-NMR spectra are shown in Figure 3.8 of Chapter 3.

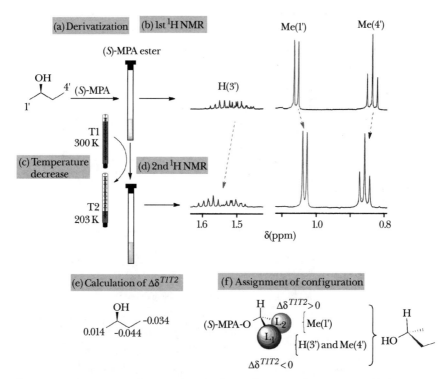

*Figure 4.6.* Main steps in the procedure for the configurational assignment of (R)-butan-2-ol, using (S)-MPA and low-temperature NMR.

## 4.2. COMPLEXATION WITH BA²⁺: MPA ESTERS OF SECONDARY ALCOHOLS

The absolute configuration of secondary alcohols (Figure 4.7) can be assigned from a single MPA derivative without resorting to changes in the probe temperature. This is accomplished by observing the changes in the spectra recorded before and after the addition of barium perchlorate to the NMR tube containing the MPA ester in MeCN-$d_3$ [42, 165].

The procedure is simple (Figure 4.1b): the sample of unknown configuration is derivatized with either (*R*)- or (*S*)-MPA, the corresponding ¹H-NMR spectrum is recorded in MeCN-$d_3$, and the signals of $L_1/L_2$ are assigned [42]. Ba(ClO$_4$)$_2$ is added, and a second spectra is recorded. The $\Delta\delta^{Ba}$ values for $L_1/L_2$ are calculated (see next paragraph), and the configuration is assigned according to the sign distribution, as shown in Figure 4.8.

In this method, a preferential complexation takes place between the metal cation and the *sp* conformer of the derivative, shifting the *sp/ap* equilibrium [37, 42] and increasing the concentration of the *sp* conformer (i.e., *sp*-Ba²⁺; Figure 4.8). The resulting effect on the spectra is very much like the one observed in the low-temperature approach (Figure 4.3), and it can be conveniently expressed as $\Delta\delta L^{Ba}$, which is defined to be $\delta L$ in the absence of Ba²⁺ minus $\delta L$ in the presence of Ba²⁺.

For instance, in the (*R*)-MPA ester of a secondary alcohol with the configuration shown in Figure 4.8a, the $L_1$ substituent is shielded in the *sp* conformer, and $L_2$ is shielded in the *ap* conformer [42]. After addition of Ba(ClO$_4$)$_2$, the population of

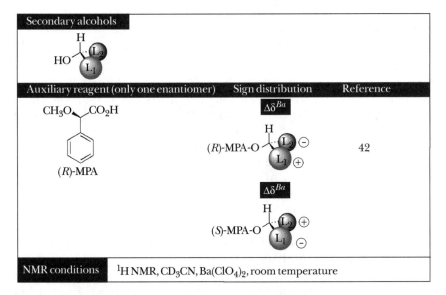

*Figure 4.7.* The NMR procedure for the assignment of secondary alcohols by complexation with Ba²⁺: CDA, sign distributions ($\Delta\delta^{Ba}$), experimental conditions, and references.

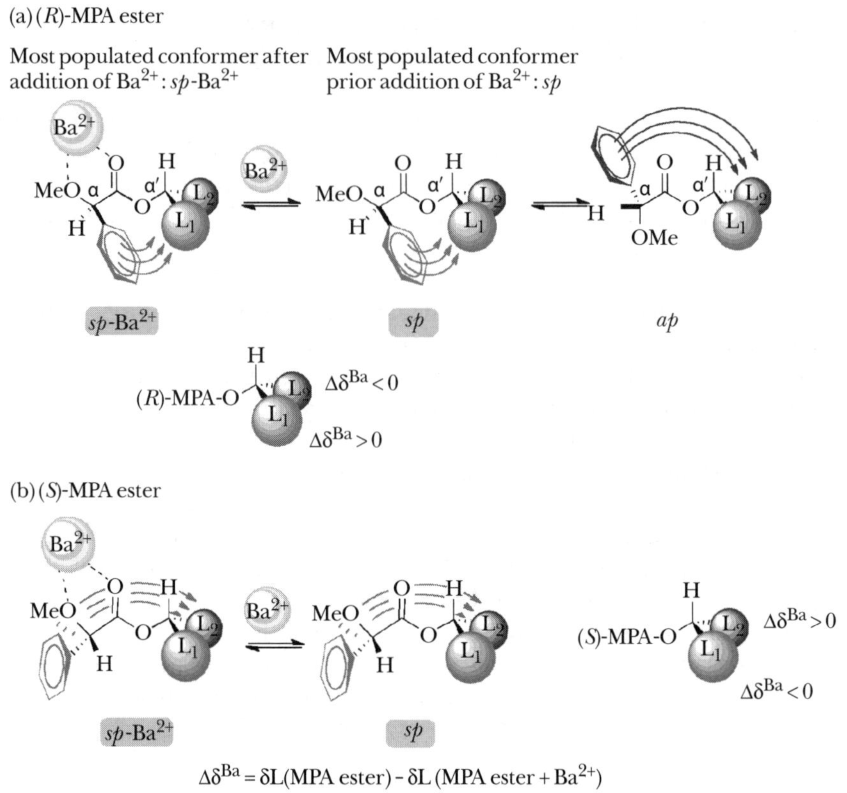

## (a) (R)-MPA ester

Most populated conformer after addition of Ba$^{2+}$: *sp*-Ba$^{2+}$

Most populated conformer prior addition of Ba$^{2+}$: *sp*

*sp*-Ba$^{2+}$ ⇌ *sp* ⇌ *ap*

(R)-MPA-O — $\Delta\delta^{Ba} < 0$ (L$_2$), $\Delta\delta^{Ba} > 0$ (L$_1$)

## (b) (S)-MPA ester

*sp*-Ba$^{2+}$ ⇌ *sp*

(S)-MPA-O — $\Delta\delta^{Ba} > 0$ (L$_2$), $\Delta\delta^{Ba} < 0$ (L$_1$)

$\Delta\delta^{Ba} = \delta L(\text{MPA ester}) - \delta L(\text{MPA ester} + \text{Ba}^{2+})$

*Figure 4.8.* Conformational equilibria and $\Delta\delta^{Ba}$ values for the MPA esters of secondary alcohols.

the *sp* conformer increases, and that of the *ap* decreases [42] (Figure 4.8a). Thus, in the presence of the salt, L$_1$ is more shielded and L$_2$ is deshielded. As a result, $\Delta\delta^{Ba}$ is positive for L$_1$ and negative for L$_2$. Naturally, when (S)-MPA is used, the opposite set of signs is obtained (negative $\Delta\delta^{Ba}$ for L$_1$, and positive for L$_2$; Figure 4.8b).

The correlation between the barium shifts and the spatial position of the L$_1$/L$_2$ groups has been validated with a large series of representative secondary alcohols [42].

Sections 4.2.1 and 4.2.2 present examples that illustrate the application of this method to the assignment of secondary alcohols.

### 4.2.1. Example 23: Assignment of the Absolute Configuration of (R)-Pentan-2-ol Using (S)-MPA

The partial $^1$H-NMR spectra of the (S)-MPA ester of an enantiomer of pentan-2-ol (MeCN-d$_3$, room temperature) in the presence and absence of Ba(ClO$_4$)$_2$ are shown in Figures 4.9b and d, respectively, and the assignment of the protons is indicated. They are located at the two branches of the asymmetric carbon, Me(1′)

Monofunctional Compounds by Single Derivatization ■ 101

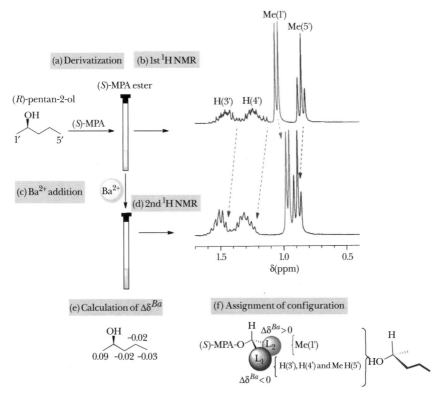

*Figure 4.9.* Main steps in the NMR procedure for the configurational assignment of (R)-pentan-2-ol, using (S)-MPA for the complexation with $Ba^{2+}$.

on one side and H(3')/H(4')/Me(5') on the other; and they are adequate for diagnosis.

Comparison of the spectra indicates that the signals from Me(1') shift to a higher field (i.e., they are shielded) upon addition of the barium salt, while those of H(3')/H(4')/Me(5') shift to a lower field (i.e., they are deshielded).

The $\Delta\delta^{Ba}$ values and signs [42] for those protons are shown in Figure 4.9e. The values corresponding to H(3')/H(4')/Me(5') are negative, while the one for Me(1') is positive. Application of the graphical model (Figure 4.8b) places the propyl chain [H(3')/H(4')/Me(5')] at the front of the tetrahedron ($L_1$), while Me(1') is placed at the back ($L_2$); this leads to the (R) absolute configuration shown in Figure 4.9f.

### 4.2.2. Example 24: Assignment of the Absolute Configuration of (R)-Pentan-2-ol Using (R)-MPA

A similar procedure using the (R) enantiomer of the auxiliary reagent [(R)-MPA] leads to the spectra shown in Figure 4.10. Comparison of the $^1$H-NMR spectra leads, as above, to the $\Delta\delta^{Ba}$ values and signs shown in Figure 4.10e: those from H(3')/H(4')/Me(5') are positive $\Delta\delta^{Ba}$, while the one from Me(1') is negative.

102 ■ The Assignment of the Absolute Configuration by NMR

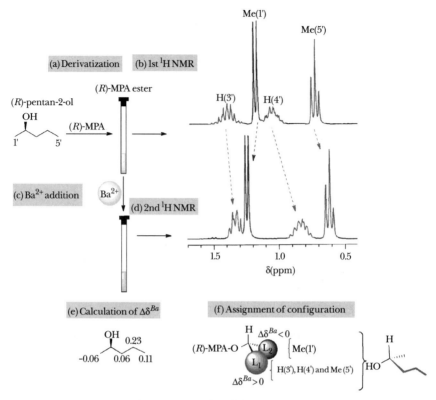

*Figure 4.10.* Main steps in the NMR procedure for the configurational assignment of (R)-pentan-2-ol, using (R)-MPA for the complexation with $Ba^{2+}$.

Application of the correlation model (Figure 4.8a and 4.10f) places the propyl chain [H(3′)/H(4′)/Me(5′)] at the front ($L_1$) and Me(1′) at the back ($L_2$) of the tetrahedron around the asymmetric carbon; this leads to the (R) absolute configuration shown in Figure 4.10f.

## ■ 4.3. COMPLEXATION WITH $BA^{2+}$: MPA AMIDES OF α-CHIRAL PRIMARY AMINES

Similar to the method used for secondary alcohols [42, 165], the addition of a barium salt can be used for the assignment of the absolute configuration of α-chiral primary amines [55, 56, 165] (Figure 4.11). The main difference is found in the conformational composition of MPA amides [52] with regard to MPA esters [37] (Figures 3.52 and 4.12): in MPA amides, *ap* is the major conformer. Therefore, the shift of the conformational equilibrium by selective complexation of the *sp* form increases the population of *sp*, the second most abundant form, not that of *ap*, the most stable one [55, 56]. In this way, the shielding effects act in exactly the same way as in alcohols with the same chirality: a positive $\Delta\delta^{Ba}$ means that the L group concerned is on the same side as the phenyl group of the CDA in the *sp* conformer [55, 56].

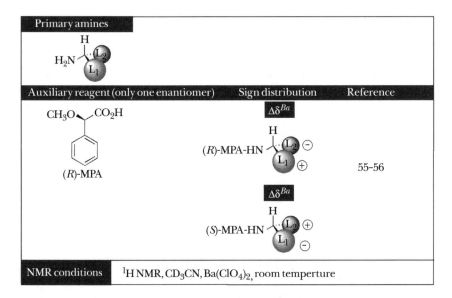

*Figure 4.11.* The NMR procedure for the assignment of α-chiral primary amines by complexation with $Ba^{2+}$: CDA, sign distributions ($\Delta\delta^{Ba}$), experimental conditions, and references.

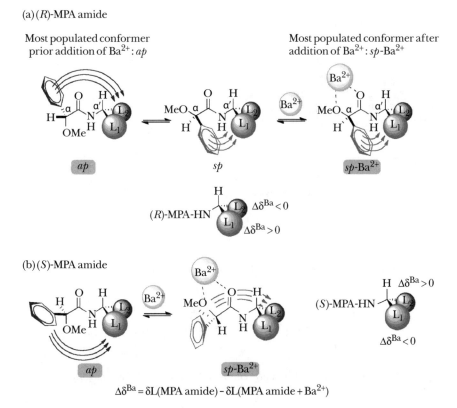

$\Delta\delta^{Ba} = \delta L(\text{MPA amide}) - \delta L(\text{MPA amide} + Ba^{2+})$

*Figure 4.12.* Conformational equilibria and $\Delta\delta^{Ba}$ values for the MPA amides of α-chiral primary amines.

In the (R)-MPA amide of an α-chiral primary amine with the configuration shown in Figure 4.12a, the $L_2$ substituent is shielded in the *ap* conformer, and the $L_1$ is shielded in the *sp* conformer. After addition of $Ba(ClO_4)_2$, the population of the *sp* conformer increases and that of the *ap* decreases (Figure 4.12a). Thus, the $L_1$ substituent shifts to a higher field (i.e., it is shielded) and $L_2$ shifts to a lower field (i.e., it is deshielded). Therefore, $\Delta\delta^{Ba}$ is positive for $L_1$ and negative for $L_2$. When, instead of an (R)-MPA amide, an (S)-MPA amide is analyzed, the opposite set of signs is obtained (i.e., negative $\Delta\delta^{Ba}$ for $L_1$ and positive for $L_2$, Figure 4.12b). This correlation between the barium shifts and the spatial position of the $L_1/L_2$ groups has been validated with a large series of representative amines and can be used for assignment [55, 56, 165].

The experimental procedure [55, 56] is similar to the one described for secondary alcohols [42] (Figure 4.1b): the sample is derivatized with either (R)- or (S)-MPA, and the corresponding $^1H$-NMR spectrum is recorded in MeCN-$d_3$. Next, a second spectrum is recorded after addition of $Ba(ClO_4)_2$. The signals of $L_1/L_2$ are assigned, and their $\Delta\delta^{Ba}$ are calculated. Finally, the configuration is assigned according to the sign distribution shown in Figure 4.12.

An example is shown in Section 4.3.1 to illustrate the application of this method.

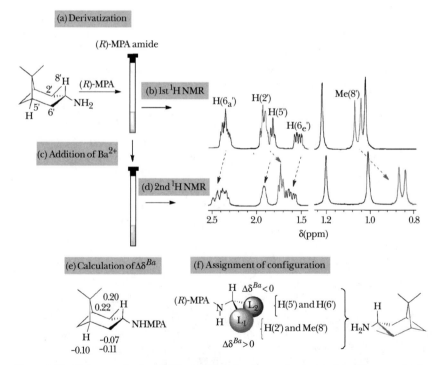

*Figure 4.13.* Main steps in the NMR procedure for the configurational assignment of (−)-isopinocampheylamine, using (R)-MPA for the complexation with $Ba^{2+}$.

### 4.3.1. Example 25: Assignment of the Absolute Configuration of (−)-Isopinocampheylamine Using (R)-MPA

Figure 4.13 shows the partial spectra of the (R)-MPA amide of (−)-isopinocampheylamine. It contains protons that are located at either side of the asymmetric carbon and serve as diagnostic signals for assignment. Comparison of the spectra in the absence and in the presence of barium perchlorate [55, 56] indicates that complexation moves the signals for H(2′)/Me(8′) to a higher field (positive $\Delta\delta^{Ba}$), while those of the H(5′)/H(6′) protons move to a lower field (negative $\Delta\delta^{Ba}$). Application of the correlation model (Figure 4.12a) places H(2′)/Me(8′) at the front of the tetrahedron ($L_1$) and the H(5′)/H(6′) protons at the back ($L_2$), leading to the stereochemistry shown in Figure 4.13f.

### ■ 4.4. ESTERIFICATION SHIFTS

A third procedure (Figure 4.1c and 4.14) can be used for the assignment of the absolute configuration of secondary alcohols [43, 165]. It does not require the use of low temperature or complexation, but it does require the use of 9-AMA as the CDA. It consists in the comparison of the chemical shifts observed for $L_1/L_2$ in the free alcohol with those for the same groups either in the (R)- or in the (S)-9-AMA ester derivative. Figure 4.14 shows the basic information, substrate, CDA, correlation model, and the NMR conditions for the assignment.

The procedure is based on the known conformational composition of 9-AMA esters [37, 39] and is analogous to that of MPA esters: sp is the main conformer,

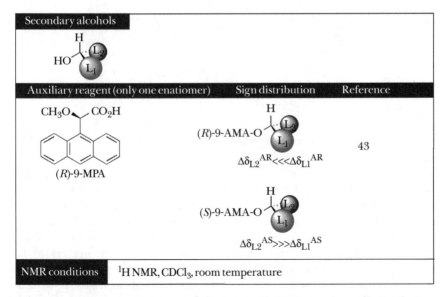

*Figure 4.14.* The assignment of secondary alcohols by the NMR procedure with esterification shifts: CDA, sign distributions ($\Delta\delta^{AR}$, $\Delta\delta^{AS}$), experimental conditions, and references.

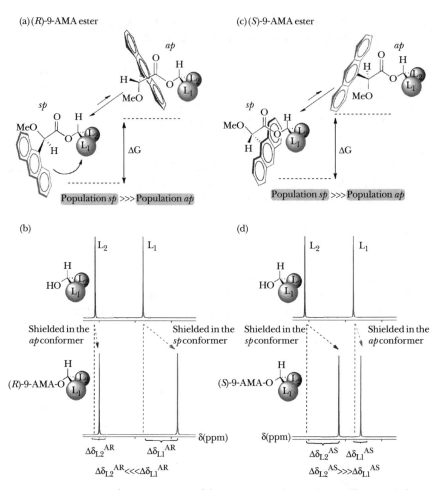

*Figure 4.15.* Conceptual representation of the NMR procedure with esterification shifts for secondary alcohols (single-derivatization procedure) making use of 9-AMA esters.

with the anthryl group generating a strong shielding effect [37, 39] (Figure 4.15). The L substituent located on the same side as the anthryl group in the *sp* conformer is very strongly shielded compared to its signal in free alcohol, while the shieldings experienced by the protons of the other L substituent are clearly smaller. In this way, the protons located on the same side of the anthryl group can be easily distinguished [43] from those on the other side by the intensity of the shielding experienced when the alcohol is derivatized with 9-AMA ("esterification shifts"). These differences are conveniently expressed through the $\Delta\delta^{AR}$ parameters for the (R)-9-AMA ester derivatives; this is defined as $\Delta\delta L^{AR}$ = $\delta L$ in the alcohol minus $\delta L$ in the (R)-9-AMA ester, or if we use (S)-9-AMA, then it is defined as $\Delta\delta L^{AS}$ = $\delta L$ in the alcohol minus $\delta L$ in the (S)-9-AMA ester. This leads to the correlation models shown in Figure 4.15. So, in the (R)-9-AMA esters, the most intensely shielded protons are found at the $L_1$ substituent (i.e., the substituent located facing the anthryl group in the *sp* conformation), while those that are only slightly shielded are located at the $L_2$ substituent. As a result, $\Delta\delta^{AR}L_2 \ll \Delta\delta^{AR}L_1$ (Figures 4.15a and b), and the reverse situation holds for (S)-9-AMA esters, where $\Delta\delta^{AS}L_2 \gg \Delta\delta^{AS}L_1$ (Figures 4.15c and d).

Two examples are shown next (Sections 4.4.1 and 4.4.2) to illustrate the application of this method [43] to the assignment of sec-alcohols.

### 4.4.1. Example 26: Assignment of the Absolute Configuration of (1R, 4S)-Hydroxycyclopent-2-en-1-yl Acetate as (R)-9-AMA Ester

Figure 4.16 shows the ¹H-NMR spectra of a cis 4-hydroxyciclopent-2-en-1-yl acetate and its (R)-9-AMA ester derivative, and it indicates the steps to be followed for the configurational assignment [43]. In this case, the signals for diagnosis are the methylene protons [H(5')] on one side of the asymmetric carbon and the vinylic protons [H(2')/H(3')] on the other. Comparison of the two spectra shows that the H(2')/H(3') resonate in the same location in the free alcohol spectrum as in the (R)-9-AMA ester spectrum ($\Delta\delta^{AR}$ is close to null: 0.01 and 0.05 ppm), while the signals for the two H(5') protons present a much larger difference (0.41 and

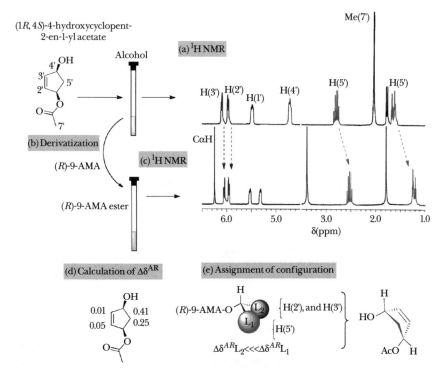

*Figure 4.16.* Main steps in the configurational assignment of (1R, 4S)-hydroxycyclopent-2-en-1-yl acetate as (R)-9-AMA ester by the NMR procedure with esterification shifts.

0.25 ppm). Application of the correlation model ($\Delta\delta^{AR}L_2 \ll \Delta\delta^{AR}L_1$; Figure 4.15) places the protons with a larger difference [H(5′), i.e., $L_1$] at the front of the tetrahedron and the protons with a smaller difference [H(2′)/H(3′), i.e., $L_2$] at the back, leading to the (4S) configuration shown in Figure 4.16e.

### 4.4.2. Example 27: Assignment of the Absolute Configuration of (1R, 4S)-Hydroxycyclopent-2-en-1-yl Acetate as (S)-9-AMA Ester

Figure 4.17 shows the data obtained when the alcohol shown in Section 4.4.1 is derivatized with the (S) enantiomer of the auxiliary 9-AMA. In this case, comparison of the $^1$H-NMR spectra from the free alcohol with that from its (S)-9-AMA ester shows that the chemical shifts of H(2′)/H(3′) clearly move upfield, while those of the two H(5′) protons remain in almost the same position ($\Delta\delta^{AS}$ is close to null; Figure 4.17d). Application of the correlation model for the (S)-9-AMA derivatives ($\Delta\delta^{AS}L_2 \gg \Delta\delta^{AS}L_1$; Figure 4.15) places the protons with larger $\Delta\delta^{AS}$ differences at the back ($L_2$) and those with the null shift at the front ($L_1$). Thus the

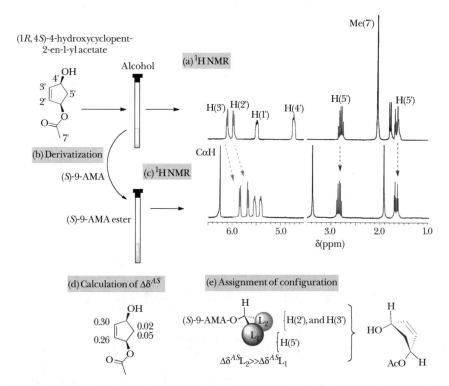

*Figure 4.17.* Main steps in the configurational assignment of (1R, 4S)-hydroxycyclopent-2-en-1-yl acetate as (S)-9-AMA ester by the NMR procedure with esterification shifts.

H(5′) protons are located at the front of the tetrahedron (L$_1$) and the H(2′)/H(3′) ones are located at the back (L$_2$), as expected.

## ■ 4.5. SUMMARY

In this chapter, we described three methods that allow the assignment of configuration of secondary alcohols [41–43, 165] and α-chiral primary amines [55, 56, 165] by $^1$H NMR and using only one derivative, either the (R)- or the (S)-ester/amide.

For secondary alcohols, there are two alternatives to doing this with MPA as the auxiliary: low-temperature NMR [41] or selective complexation [42] with Ba$^{2+}$. In a third procedure, it is necessary to use 9-AMA as the auxiliary [43]. For α-chiral primary amines, the only single-derivatization method available is based on selective complexation of the MPA amides [55, 56] with Ba$^{2+}$.

The low-temperature procedure requires MPA as the CDA, and it is applicable to secondary alcohols [41]. However, instead of comparing the spectra of the (R)- and the (S)-MPA esters and evaluating the signs of $\Delta\delta^{RS}$ for L$_1$/L$_2$, it is necessary to consider the $^1$H-NMR spectra of only one of those derivatives, but at two different temperatures (i.e., 300K and 170K) and to calculate $\Delta\delta^{T1T2}$ for the L$_1$/L$_2$ substituents. This approach can also be used for the assignment of the absolute stereochemistry of ketone cyanohydrins [49] by low-temperature NMR of the corresponding MPA ester.

It is important to point out that, since MTPA esters present a more complex conformational composition, when MTPA is used instead of MPA, the $\Delta\delta^{T1T2}$ values are smaller, and this can result in inhomogeneity of the sign distribution and misassignments [166, 167].

An alternative method that does not require the use of low-temperature NMR and is applicable to both amines [55, 56] and alcohols [42] (complexation with Ba$^{2+}$ procedure) consists in the comparison of the spectrum of the MPA ester (or MPA amide) registered in MeCN-d$_3$, and examination of the changes produced on the $\Delta\delta^{Ba}$ signs of the L$_1$/L$_2$ substituents by the addition of barium perchlorate.

The effect of the barium salt is complexation with the carbonyl and methoxy oxygens of the MPA ester/amide, and thus caution should be taken when the substrate possesses 1,4-ketoester or similar moieties that are able to form a complex with the barium cation and compete with the MPA unit. In this case, the configuration derived from this method may well be erroneous [185].

There is a third possibility for the single-derivative assignment of secondary alcohols [43] (esterification shifts procedure), using 9-AMA as the CDA. In this case, the two spectra to be compared are those from the chiral alcohol and from the corresponding 9-AMA ester, both registered in chloroform at room temperature.

These single-derivatization procedures are of especial interest in those cases where the amount of available substrate is too small for two derivatives to be prepared. If this is not a problem, comparison of the spectra from the (R)- and the (S)-CDA derivatives is recommended.

Figures 3.15 and 3.58 show a selection of secondary alcohols and primary amines with known absolute configurations that have been used to validate the procedures described in this chapter.

Other examples of the application of the low-temperature [168–176], complexation with $Ba^{2+}$ procedure to secondary alcohols [177–182] and amines [183, 184], and applications of esterification shifts [186] can be found in the literature.

# 5 Assignment of the Absolute Configuration of Polyfunctional Compounds

## ■ 5.1. SEC/SEC-1,2- AND SEC/SEC-1,N-DIOLS

### 5.1.1. Double-Derivatization Methods: MPA, 9-AMA, and MTPA

From a practical point of view, the assignment of the absolute configuration of sec/sec 1,2- and 1,n-diols does not require the separate derivatization (two different steps with the CDA of choice) of each one of the two hydroxyl groups present in the substrate; on the contrary, it can be carried out by simultaneous derivatization of the two hydroxyls (a single step), leading to the corresponding bis-(R)- and bis-(S)-CDA esters [13, 59–61]. The most used CDAs are 9-AMA and MPA [59, 60], although 1-NMA, 2-NMA, and MTPA are also appropriate [59, 60] (Figure 5.1).

This assignment has an important difference compared to that of monofunctionalized compounds [15]; this is due to the presence in the bis-(R)- and bis-(S)-derivatives of two CDA units that produce distributions of $\Delta\delta^{RS}$ and $\Delta\delta^{SR}$ signs that do not follow the trends found in monoderivatized compounds [13, 15, 82]. This means that the NMR spectra of the bis-CDA derivatives cannot be interpreted as if they had originated from two isolated mono-CDA derivatives [82]. Thus, the correlations described for secondary alcohols [35–39] cannot be applied to diols [59–61] because the chemical shifts and $\Delta\delta^{RS}$ values result from the combination of the anisotropic effects—usually shielding—from the two CDA units and not from a single unit, as happens with monoalcohols.

A result of the combination of aromatic shielding effects [59, 60] in diols is that the diagnostic protons/signals for assignment are not always the same as in isolated monoalcohols (i.e., $L_1/L_2$). For instance, in acyclic syn-1,2-diols, the diagnostic signals [59, 60] are those corresponding to the protons at the alpha positions of the OH groups (i.e., the hydrogens linked directly to the asymmetric carbons) $H\alpha(R_1)$ and $H\alpha(R_2)$ exclusively. On the other hand, in acyclic anti-1,2-diols, the diagnostic signals are from $H\alpha(R_1)/H\alpha(R_2)$ together with those from $R_1$ and $R_2$ (Figure 5.2a).

As in the case of monofunctional compounds, the assignment consists [13, 59, 60] in the preparation of two bis-CDA derivatives from the two enantiomers of the chosen CDA, followed by comparison of the corresponding NMR spectra and calculation of the $\Delta\delta^{RS}$ (or $\Delta\delta^{SR}$ in the case of MTPA) signs for $H\alpha(R_1)$, $R_1$, $H\alpha(R_2)$, and $R_2$. The $\Delta\delta$ parameters show sign distributions that are specific for each of the four possible configurations of the sec/sec-1,2-diols (types A–D, Figure 5.2a), allowing their use for assignment [59, 60].

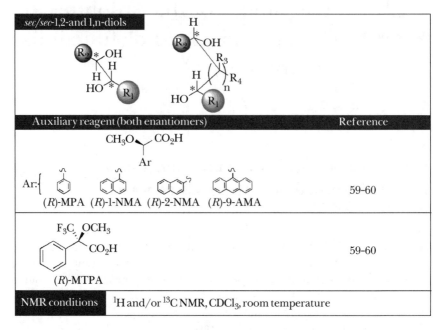

*Figure 5.1.* The assignment of *sec/sec*-1,2- and *sec/sec*-1,n-diols by double-derivatization methods at a glance: CDAs, experimental conditions, and references.

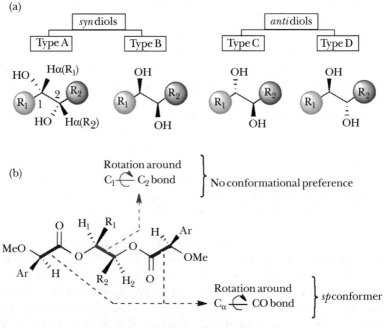

*Figure 5.2.* Classification of 1,2-diols according to structural types and the main bonds involved in the conformational equilibria.

The origin of these correlations can be traced, as in monoderivatized compounds, to the main conformers of the bis-(R)- and bis-(S)-CDA esters and the orientation of the aryl groups of the two CDA units relative to the rest of the molecule [59, 60].

In the case of bis-MPA and bis-9-AMA esters, conformational studies [60] show that the two MPA auxiliaries (or other AMAAs, such as 9-AMA), have the same conformational preferences around the Cα-CO bond as in the mono-MPA esters of sec-alcohols [37]; this is an sp/ap equilibrium, sp being the major conformer (Figure 5.2b). For its part, rotation around the C(1)-C(2) bond does not present any significant conformational preference [60]. Therefore, the chemical shifts observed for the stereoisomers of a sec/sec-1,2-diol (types A-D, Figure 5.2a) can be explained as having originated due to structures similar to that shown in Figure 5.2b, with the two MPA units in the sp conformation [60].

For example, if the two AMAA (i.e., CDA) units of a type A 1,2-diol (Figure 5.2a) adopt the sp conformation, the $R_1$ and $R_2$ substituents are located under their shielding cones both in the bis-(R)-AMAA ester as well as in the bis-(S)-AMAA ester (Figure 5.3a). As we cannot know the relative intensity of each shielding, we are not able to predict the resulting $\Delta\delta^{RS}$ signs for $R_1$ and $R_2$ in that stereoisomer. Thus, no correlation between those signs and the stereochemistry can be established [60].

On the other hand, $H\alpha(R_1)$ and $H\alpha(R_2)$ are under the shielding cones in the bis-(S)-AMAA ester but not in the bis-(R)-AMAA ester (Figure 5.3a). Therefore, the chemical shifts of $H\alpha(R_1)$ and $H\alpha(R_2)$ are diagnostic for that stereochemistry and can be expressed through their positive $\Delta\delta^{RS}$ value [60] (Figure 5.3a).

Similarly, in a type B diol (Figure 5.3b), $R_1$ and $R_2$ cannot be used as diagnostic signals, because both groups are again shielded in the two bis-AMAA esters [60]. Fortunately, $H\alpha(R_1)$ and $H\alpha(R_2)$ are sensitive to the stereochemistry, and both are more shielded in the bis-(R)- than in the bis-(S)-AMAA ester. This produces the diagnostic signals and results in a negative $\Delta\delta^{RS}$ for both hydrogens (Figure 5.3b).

For type C diols (Figure 5.3c), $R_1$ and $H\alpha(R_1)$ are more shielded in the bis-(R)-AMAA ester, while $R_2$ and $H\alpha(R_2)$ are more shielded in the bis-(S)-AMAA ester. Therefore, $R_1$, $R_2$, $H\alpha(R_1)$ and $H\alpha(R_2)$ generate diagnostic signals. The $\Delta\delta^{RS}$ values for $R_1$ and $H\alpha(R_1)$ are negative in that stereochemistry, while those for $R_2$ and $H\alpha(R_2)$ are positive [60].

Finally, for type D diols (Figure 5.3d), $\Delta\delta^{RS}$ for $R_2$ and $H\alpha(R_2)$ are negative because they are more shielded in the bis-(R)- than in the bis-(S)-AMAA ester, while those due to $R_1$ and $H\alpha(R_1)$ are positive, because they are more shielded in the bis-(S)- than in the bis-(R)-AMAA ester [60].

These results are summarized for the four stereoisomers of sec/sec-1,2-diols in Figures 5.4a and b.

Application of the same reasoning to acyclic sec/sec-1,n-diols allows us to determine which signals are diagnostic and to determine the correlation between their $\Delta\delta^{RS}$ signs and their stereochemistry [13, 59, 60]. It is necessary to point out that in sec/sec-1,n-diols, besides $R_1$, $R_2$, $H\alpha(R_1)$, and $H\alpha(R_2)$, the shifts of the protons placed between the chiral carbons may also be diagnostic (i.e., Types A and B). These results are shown in Figures 5.4c–f.

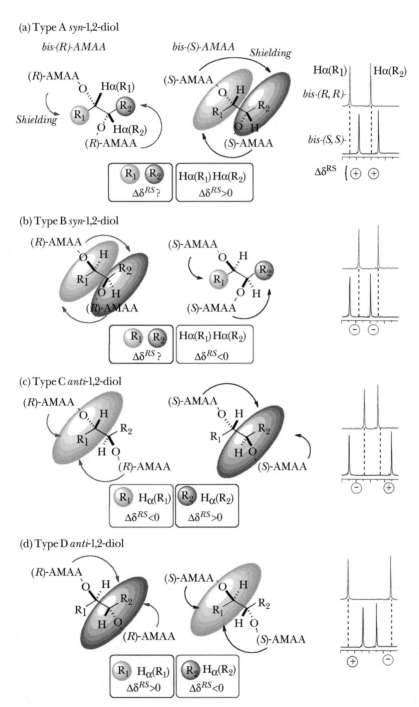

*Figure 5.3.* Shielding effects and $\Delta\delta^{RS}$ values for the bis-AMAA esters of the four structural types of *sec/sec*-1,2-diols. Conceptual representations on the origin of the $\Delta\delta^{RS}$ signs corresponding to $H\alpha(R_1)$ and $H\alpha(R_2)$ from the $^1$H-NMR spectra are also shown.

Polyfunctional Compounds ■ 115

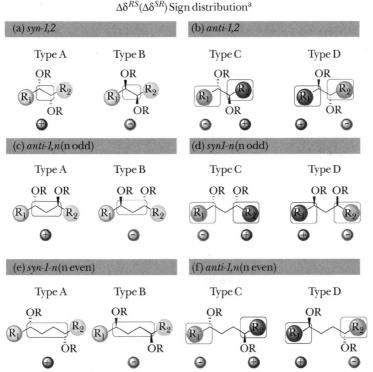

$^a$Sign distribution: $\Delta\delta^{RS}$, CDA = MPA, 9-AMA, 1-NMA, 2-NMA (AMAAs); $\Delta\delta^{SR}$, CDA = MTPA

Figure 5.4. $\Delta\delta^{RS}$ and $\Delta\delta^{SR}$ sign distributions for acyclic sec/sec-1,2-diols and sec/sec-1,n-diols.

As the main conformational characteristics of 2-NMA and 1-NMA are similar to those of MPA and 9-AMA, the correlations are the same between the sign distributions of $\Delta\delta^{RS}$ and the stereochemistry when 2-NMA or 1-NMA are used instead of MPA and 9-AMA [13, 59, 60].

When MTPA is utilized, the general patterns represented in Figure 5.4 for bis-AMAA esters are also applicable to bis-MTPA esters by simply replacing $\Delta\delta^{RS}$ (AMAA) by $\Delta\delta^{SR}$ (MTPA) values [13, 59, 60]. This is due to the conformational differences observed between bis-AMAA and bis-MTPA esters: as shown in Section 3.1 (secondary alcohols), if substituent $L_1$ is shielded in the (R)-AMAA ester of a secondary alcohol, $L_2$ is unaffected; the situation is reversed in the corresponding (R)-MTPA ester (i.e., $L_2$ is shielded, and $L_1$ is unaffected) [38]. As a consequence, the signs of $\Delta\delta^{RS}$ obtained with MTPA are the opposite of those obtained with AMAAs. But as the chemical shift differences are calculated with MTPA in the form of $\Delta\delta^{SR}$ instead of $\Delta\delta^{RS}$, the sign patterns are the same in both cases. The same reasoning applies to sec/sec-diols [13, 59, 60].

Finally, as is true for other cases, $^{13}$C-NMR chemical shifts can also be used for assignment [72] by treating the signs of $^{13}$C $\Delta\delta^{RS}$ in exactly the same way as we treat those from $^1$H NMR.

## 5.1.2. Example 28: Assignment of the Absolute Configuration of Heptane-2,3-Diol (*Syn*)

A sample of heptane-2,3-diol of unknown absolute configuration is separately derivatized with two equivalents of (*R*)-MPA and two equivalents of (*S*)-MPA, and the $^1$H-NMR spectra of the resulting bis-MPA esters are registered (Figures 5.5a and b).

The first step consists in determining if the diol is *syn* or *anti*. As shown above (Section 5.1.1), the diagnostic signals for *syn* diols are those from $H\alpha(R_1)$ and $H\alpha(R_2)$, and therefore the assignment of *syn* diols requires the identification of those methine protons only. However, in the case of *anti* diols, the identification of those methines [$H\alpha(R_1)$ and $H\alpha(R_2)$] are necessary, as well as the signals from $R_1$ and $R_2$. Therefore, from a practical point of view, the easiest approach is to identify in the spectra the signals from the methine protons first, and then to see if the diol is *syn* (same $\Delta\delta^{RS}$ signs) or *anti* (opposite $\Delta\delta^{RS}$ signs).

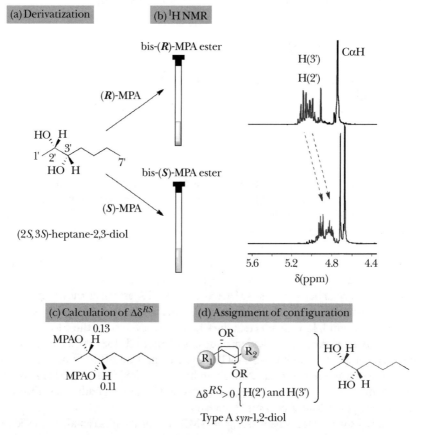

*Figure 5.5.* Main steps in the configurational assignment of heptane-2,3-diol (*syn*) using MPA.

In this way, the signals corresponding to $H\alpha(R_1)$ and $H\alpha(R_2)$ in the 4.8–5.1 ppm region of the spectra are assigned first (Figure 5.5b), and their $\Delta\delta^{RS}$ values are calculated (Figure 5.5c). In this example, the experimental $\Delta\delta^{RS}$ values have the same sign [both positive; 0.13 and 0.11 ppm for $H\alpha(2')$ and $H\alpha(3')$, respectively], which indicates that the diol is *syn*, and there is no need to assign the $R_1/R_2$ signals in the spectra. Next, the distinction between the two enantiomeric *syn* diols (types A and B) arises from the positive $\Delta\delta^{RS}$ signs for both $H\alpha(R_1)$ and $H\alpha(R_2)$, which are characteristic of a type-A diol (Figure 5.4a). Therefore, this compound is a type A *syn*-1,2-diol (Figure 5.5d), and its absolute configuration is (2S, 3S).

If we consider the main *sp* conformations [60] of the bis-(R)- and bis-(S)-MPA esters of (2S, 3S)-heptane-2,3-diol (Figure 5.6), the different chemical shifts observed in the spectra can be easily explained and justified. In these conformations, the $R_1$ and $R_2$ substituents [Me(1') and H(4'-6')/Me(7'), respectively] are located under the shielding cones of the two MPA units in both the bis-MPA esters (Figure 5.6). Therefore, $R_1$ and $R_2$ are discarded as diagnostic signals, because it is not possible to predict which one is more shielded than the other nor is it possible to predict the differences in their chemical shifts (i.e., their $\Delta\delta^{RS}$ signs). However, $H\alpha(R_1)$ and $H\alpha(R_2)$ [H(2') and H(3'), respectively] are more shielded in the bis-(S)-MPA ester than in the bis-(R)-MPA-ester (Figure 5.6), leading to the diagnostic positive $\Delta\delta^{RS}$ signs that are experimentally observed.

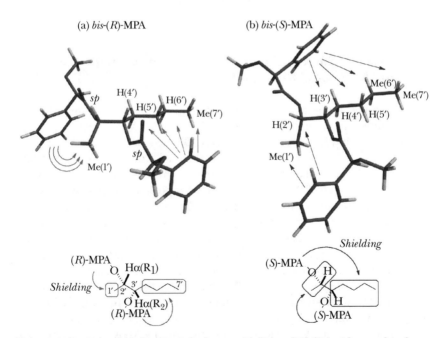

*Figure 5.6.* Position of the diagnostic hydrogens $H\alpha(R_1)$ and $H\alpha(R_2)$ with regard to the phenyl groups in the *sp* conformers of the two bis-MPA esters of (2S, 3S)-heptane-2,3-diol.

118 ■ The Assignment of the Absolute Configuration by NMR

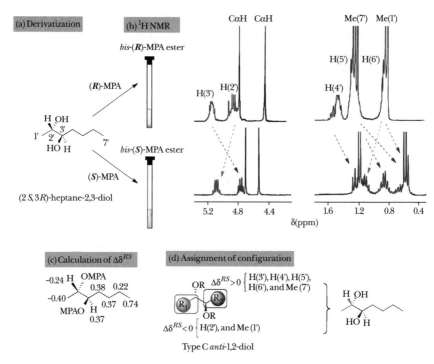

*Figure 5.7.* Main steps in the configurational assignment of heptane-2,3-diol (*anti*) using MPA.

### 5.1.3. Example 29: Assignment of the Absolute Configuration of Heptane-2,3-Diol (*Anti*)

In this example, a pure enantiomer of a heptane-2,3-diol of unknown absolute configuration is separately derivatized with (*R*)- and (*S*)-MPA, and the $^1$H-NMR spectra of the bis-MPA esters are registered (Figures 5.7a and b). The signals corresponding to $H\alpha(R_1)$ and $H\alpha(R_2)$ are assigned, and their $\Delta\delta^{RS}$ are calculated (Figure 5.7c). Unlike the previous example, the $\Delta\delta^{RS}$ values obtained for the two methines [$H\alpha(R_1)$ and $H\alpha(R_2)$] are of opposite signs (−0.24 and +0.37 ppm, respectively), indicating that the diol belongs to the *anti* class, and therefore requiring the identification of the NMR signals for $R_1$ and $R_2$ and using their $\Delta\delta^{RS}$ for the assignment (Figure 5.7c). Thus, $H\alpha(R_1)$ and $R_1$ [i.e., H(2′) and Me(1′)] have negative $\Delta\delta^{RS}$ values, while $H\alpha(R_2)$ and $R_2$ [i.e., H(3′) and H(4′-6′)/Me(7′)] have positive ones (Figure 5.7c). Comparison of the experimental $\Delta\delta^{RS}$ signs with those shown in Figure 5.4b indicate that the compound is a type C *anti*-1,2-diol with the structure shown in Figure 5.7d and with the (2*S*, 3*R*) absolute configuration.

Analysis of the shielding effects transmitted by the MPA units in the most representative conformation (*sp*), explains the chemical shifts and the $\Delta\delta^{RS}$ sign distribution that is obtained (Figure 5.8).

Polyfunctional Compounds ▪ 119

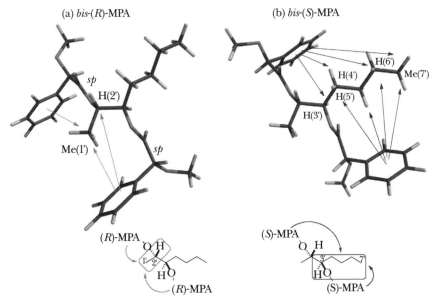

*Figure 5.8.* Position of the diagnostic hydrogens Hα($R_1$)/$R_1$ and Hα($R_2$)/$R_2$ with regard to the phenyl groups in the *sp* conformers of the two bis-MPA esters of (2S, 3R)-heptane-2,3-diol.

Thus, in the bis-(R)-MPA ester, the $R_1$ and Hα($R_1$) [Me(1′) and H(2′)] are doubly shielded by the two MPA units (Figure 5.8), while in the bis-(S)-MPA ester, those protons are unaffected, resulting in the $\Delta\delta^{RS}$ difference being negative. On the other hand, $R_2$ and Hα($R_2$) [H(4′-6′)/Me(7′) and H(3′)] are doubly shielded in the bis-(S)-MPA ester and unaffected in the bis-(R)-MPA ester, leading to a positive $\Delta\delta^{RS}$ (Figure 5.8).

### 5.1.4. Example 30: Assignment of the Absolute Configuration of 1,4-Bis-O-(4-Chlorobenzyloxy)-D-Threitol (*Syn*) Using $^{13}$C NMR

As previously mentioned (e.g., Section 5.1.1), $^{13}$C-NMR chemical shifts can also be used for the assignment [72] by treating the signs of $^{13}$C $\Delta\delta^{RS}$ in exactly the same way as we used those due to $^{1}$H NMR. We show the $^{13}$C NMR and $\Delta\delta^{RS}$ of the bis-MPA ester of 1,4-bis-O-(4-chlorobenzyloxy)-D-threitol as an example (Figure 5.9b).

Figure 5.9c shows negative $\Delta\delta^{RS}$ signs for carbons C(2′) and C(3′) (−0.17 ppm), indicating that the diol is *syn* (same $\Delta\delta^{RS}$ signs), and there is thus no need to assign the $R_1/R_2$ signals in the spectra. A negative distribution of $\Delta\delta^{RS}$ signs is coincident with the one found for type B *syn* diols (Figure 5.3b and 5.4a). Therefore, the absolute configuration of the diol is (2R, 3R).

This result is explained by analysis of the shielding effects transmitted to the rest of the molecule by the MPA units placed in the *sp* conformation.

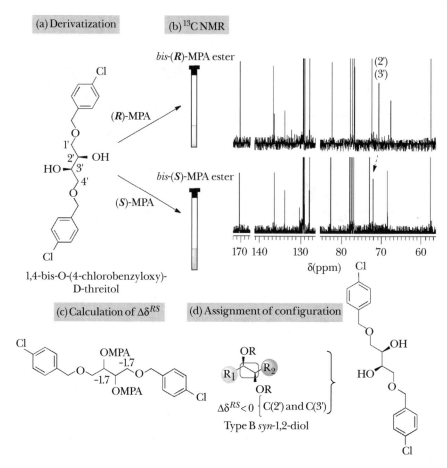

*Figure 5.9.* Main steps in the configurational assignment of 1,4-bis-O-(4-chlorobenzyloxy)-D-threitol (*syn*) using $^{13}$C NMR.

In the bis-(*R*)-MPA ester, all of the carbons in the substrate are located under the shielding cone of the two MPA units (Figure 5.10a), and in the bis-(*S*)-MPA ester, all of the carbons except C(2′) and C(3′) are shielded (Figure 5.10b). Therefore, C(2′) and C(3′) are more shielded in the bis-(*R*)-MPA ester than in the bis-(*S*)-MPA-ester (Figure 5.10), leading to the negative $\Delta\delta^{RS}$ signs that are experimentally observed.

### 5.1.5. Single-Derivatization Methods: MPA

Some of the single-derivatization methods described for monofunctional substrates have also been shown to be useful in certain types of difunctional compounds [41, 165]. This is the case for *sec/sec*-1,2 diols, which can be assigned using the low-temperature procedure and taking the chemical shifts of $R_1$ and $R_2$ as the diagnostic signals [13, 61] (Figure 5.11).

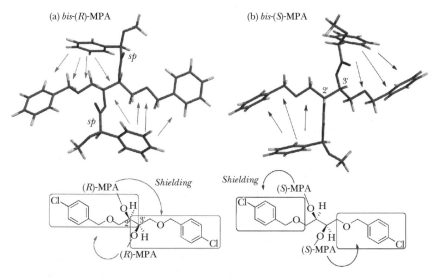

*Figure 5.10.* Position of the diagnostic carbons with regard to the phenyl groups in the *sp* conformers of the two bis-MPA esters of 1,4-bis-O-(4-chlorobenzyloxy)-D-threitol.

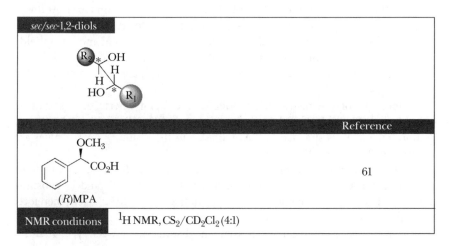

*Figure 5.11.* The assignment of *sec/sec*-1,2-diols by single-derivatization methods at a glance: CDA, experimental conditions, and references.

This procedure [61] requires the preparation of only one bis-MPA ester [either the bis-(R) or the bis-(S)] and the comparison of its NMR spectra in $CS_2/CD_2Cl_2$ (4:1) at room temperature (298 K) and at a lower temperature (e.g., 213 K). The differences in chemical shifts of $R_1$ and $R_2$ (i.e., the difference between the chemical shifts at room temperature minus those at the lower temperature) are expressed as $\Delta\delta^{T1T2}$, for which the signs are characteristic of the stereochemistry [61]. In practice, however, the correlation between the $\Delta\delta^{T1T2}$ signs and the stereochemistry is limited, because it is not sufficient to distinguish between the two enantiomers of the *syn* pair [61].

In this way, the method allows identification of the *syn* pair, but it is unable to distinguish between type A and type B. It also allows identification of the two enantiomers of the *anti* pair (types C and D). Overall, three out of the four possible structures can be distinguished [61].

As in the case of secondary alcohols, this low-temperature procedure is based on the controlled shift of the *ap/sp* conformational equilibrium in the two MPA units—the changes on the rotation of the C(1)-C(2) bond caused by the temperature variations do not influence the chemical shifts of $R_1$ and $R_2$—and the evolution of the signals from $R_1$ and $R_2$.

A decrease in temperature increases the population of *sp* (the most stable conformer) and reduces that of *ap* (the least stable one). Therefore, the signals from the $R_1/R_2$ groups that are under the shielding cone in the *sp* conformation will show a greater shielding at lower temperature as a response to the increase in the number of molecules in the equilibrium that have those $R_1/R_2$ shielded, while those due to the $R_1/R_2$ groups that are under the shielding cone in the *ap* conformer will show a reduced shielding at lower temperature [41, 61].

Thus, in the *sp* conformations of the bis-(*R*)-MPA ester of a type C *anti*-1,2-diol (Figure 5.12a), the two MPA units simultaneously shield $R_1$, whereas those two MPA units shield $R_2$ in the *ap* conformations [61]. At lower temperatures, the equilibrium shifts towards the *sp* form, and therefore the number of molecules with shielded $R_1$ increases (the signal moves upfield) while the number of those with shielded $R_2$ decreases (the signal moves downfield). As a result, the chemical shifts differences (i.e., $\Delta\delta^{T1T2}$) are positive for $R_1$ and negative for $R_2$.

In the case of the bis-(*R*)-MPA ester of the enantiomeric diol, a type D *anti*-1,2-diol (Figure 5.12b), the behavior is reversed (i.e., $R_2$ is shielded in the *sp* conformer, and $R_1$ is shielded in the *ap* one), and the resulting $\Delta\delta^{T1T2}$ values are negative for $R_1$ and positive for $R_2$.

For type A and type B *syn*-1,2-diols, the evolution of the spectra with temperature is analogous for both stereochemistries [61]. Both $R_1$ and $R_2$ substituents are shielded in both conformers (*sp/ap*) of the bis-(*R*)-MPA ester. Therefore, when the temperature is decreased, the signals from the two groups shift upfield (i.e., positive $\Delta\delta^{T1T2}$), rendering impossible the distinction between types A and B (Figures 5.12c and d).

Finally, if (*S*)-MPA were used instead of (*R*)-MPA to derivatize the diol, similar reasoning leads to a reversal of the $\Delta\delta^{T1T2}$ sign distribution [61]. These results are summarized in Figure 5.13.

It is necessary to point out that if the NMR data obtained with one bis-MPA ester showed that the diol is *syn* (same signs for the $R_1/R_2$ groups), the complete assignment requires the preparation of the other MPA derivative and comparison of the $\Delta\delta^{RS}$ signs for the $H\alpha(R_1)/H\alpha(R_2)$ and $R_1/R_2$ substituents, as shown in Section 5.1.1.

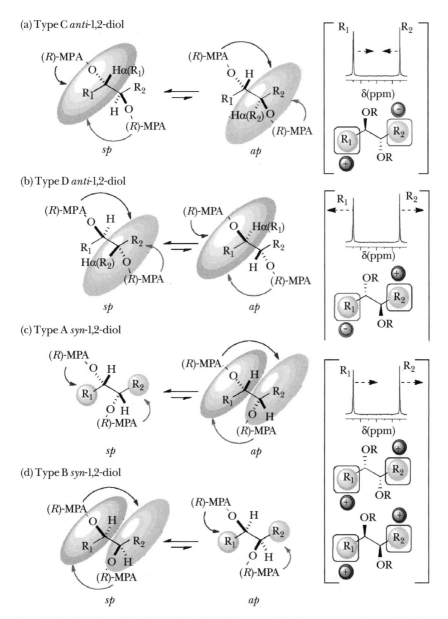

*Figure 5.12.* Shielding effects for the bis-(R)-MPA esters of the four structural types of *sec/sec*-1,2-diols. Conceptual representations on the origin of the $\Delta\delta^{RS}$ signs corresponding to $R_1$ and $R_2$ from the $^1$H-NMR spectra registered at two different temperatures are also shown.

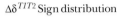
$\Delta\delta^{T1T2}$ Sign distribution

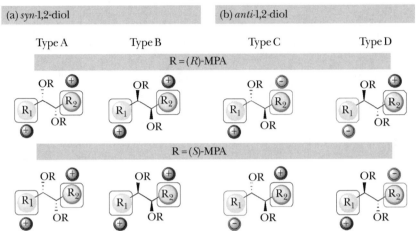

*Figure 5.13.* $\Delta\delta^{T1T2}$ sign distributions for the bis-MPA esters of *sec/sec*-1,2-diols by the low-temperature procedure.

### 5.1.6. Example 31: Assignment of the Absolute Configuration of a Pure Isomer of 3,4-Dihydroxy-5-Methylhexan-2-One

Step-by-step application of the procedure [61] for the assignment of the absolute configuration of a stereoisomer of 3,4-dihydroxy-5-methylhexan-2-one derivatized with (R)-MPA, including the identification of the spectral signals, is shown schematically in Figure 5.14.

Comparison of the spectra registered at 298 K and 183 K shows that: (a) H(5')/Me(6')/Me(7') shifts upfield at lower temperatures, and (b) Me(1') shifts downfield at lower temperatures (Figures 5.14b–d). This indicates that the substrate is an *anti*-1,2-diol, and therefore its absolute configuration can be assigned by this method. Calculation of chemical shifts differences (Figure 5.14e) shows positive $\Delta\delta^{T1T2}$ values for H(5')/Me (6')/Me(7') and a negative value for Me(1').

Comparison with the sign distributions depicted in Figure 5.13b for bis-(R)-MPA esters shows that these are the same as those of the type C *anti*-1,2-diol structure, where H(5')/Me (6')/Me(7') and Me(1') are located in the places of $R_1$ and $R_2$, respectively, leading to the (3S, 4S) configuration (Figure 5.14f).

### 5.1.7. Example 32: Assignment of the Absolute Configuration of Another Isomer of 3,4-Dihydroxy-5-Methylhexan-2-One

A sample of another pure isomer of 3,4-dihydroxy-5-methylhexan-2-one is derivatized with (R)-MPA and the spectra are registered at 298 K and 183 K, and the two are compared. The evolution of the spectra with the temperature shows: (a) H(5')/M/(6')/Me(7') are shifted downfield at lower temperatures, and

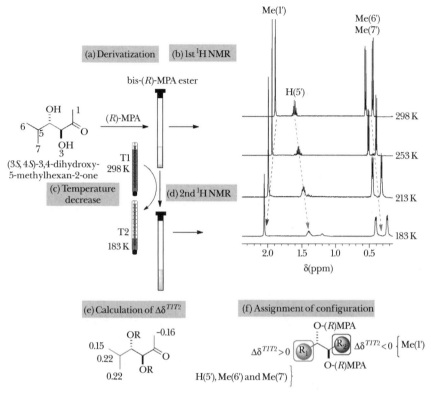

*Figure 5.14.* Main steps in the configurational assignment of a stereoisomer of 3,4-dihydroxy-5-methylhexan-2-one by the low-temperature procedure using (R)-MPA.

(b) Me(1′) is shifted upfield at lower temperatures (Figure 5.15b–d), indication that the sample is an *anti*-1,2-diol. Calculation of the chemical shifts differences shows negative $\Delta\delta^{T1T2}$ values for H(5′)/Me(6′)/Me(7′) and positive ones for Me(1′) (Figure 5.15e). Comparison of these $\Delta\delta^{T1T2}$ signs with the sign distributions shown in Figure 5.13b for bis-(R)-MPA esters shows that these match those of type D *anti*-1,2-diols, where H(5′)/Me(6′)/Me(7′) occupy the place of $R_1$ and Me(1′) is in the place of $R_2$ (Figure 5.15f); this corresponds to a (3R, 4R) absolute configuration.

### 5.1.8. Summary

The absolute configuration of the two asymmetric carbons in a *sec/sec*-1,n-diol can be simultaneously determined by double derivatization [13, 59–61] (Section 5.1.1), that is to say, by comparison of the ¹H NMR of the bis-(R)- and the bis-(S)-derivatives formed from several AMAAs. This includes MPA and its analogues with different aryl systems replacing the phenyl group, and MTPA. In all cases, the two hydroxyl groups of the diol are derivatized at the same time, and the signals

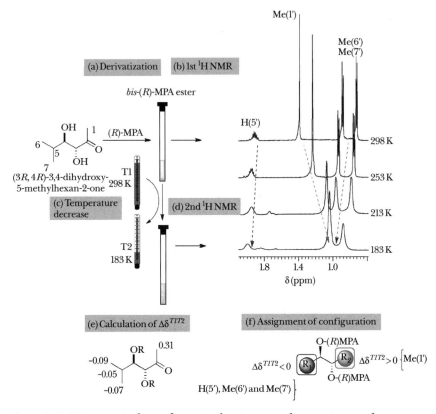

*Figure 5.15.* Main steps in the configurational assignment of a stereoisomer of 3,4-dihydroxy-5-methylhexan-2-one by the low-temperature procedure using (R)-MPA.

for diagnosis are obtained from the methine protons that are directly linked to the asymmetric carbons Hα(R$_1$) and Hα(R$_2$), which present a distinctive distribution of $\Delta\delta^{RS}$ signs for each isomer. Once the $\Delta\delta^{RS}$ signs due to the methines have identified a particular stereoisomer, the $\Delta\delta^{RS}$ signs from R$_1$ and R$_2$ can be used to confirm the assignment in case of *anti*-1,2-diols.

As happens with monoalcohols, the substitution of the phenyl ring of MPA by larger aromatic systems produces both stronger shielding effects and larger $\Delta\delta^{RS}$ values, which can be necessary for particular substrates. In most cases, the commercial availability of MPA makes it the reagent of choice [60]. If MTPA is going to be used, the reader should pay attention to the considerations about the preparation of the MTPA esters and the use of $\Delta\delta^{SR}$ instead of $\Delta\delta^{RS}$ that were mentioned in Chapter 3, Section 3.1.8.

The assignment by double derivatization can also be carried out by $^{13}$C NMR of the same derivatives [72] and then applying the same correlation models between the stereochemistry and the $\Delta\delta^{RS}$ signs; however, we note that $^{13}$C $\Delta\delta^{RS}$ values are often too small to provide a safe assignment. Finally, the assignment of absolute configuration of *sec/sec*-1,n-diols can also be carried out by $^1$H NMR of a

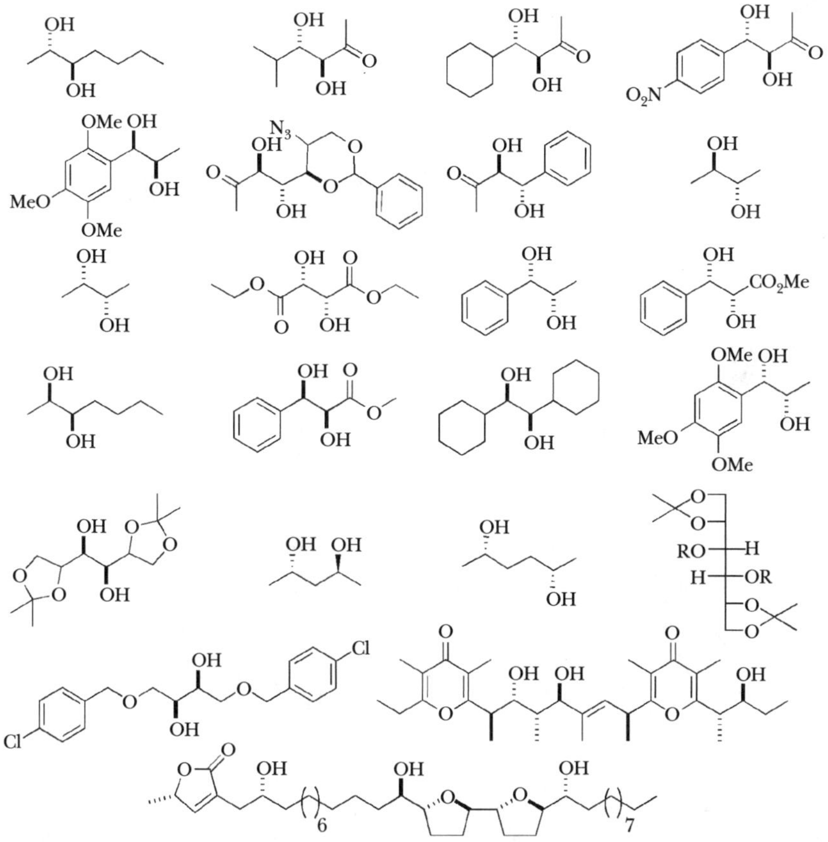

*Figure 5.16.* Selection of *sec/sec*-diols with known absolute configuration that have been used to validate the procedures described in this chapter.

single derivative [13, 61, 165] (low-temperature method; Section 5.1.5), either the bis-(*R*)- or the bis-(*S*)-MPA ester, but with some limitations, because although the two enantiomers of the *anti* pair can be identified, it is impossible to differentiate between the two enantiomers that form the *syn* pair. This means that only three (the two *anti* enantiomers and the *syn* pair) of the four stereoisomers can be distinguished by this procedure.

Figure 5.16 shows a selection of diols of known configuration that have been used to validate all of these methods [59–61]. Other examples of applications of this methodology can be found in the literature [187–213].

## ■ 5.2. SEC/SEC-1,2-AMINO ALCOHOLS

### 5.2.1. Double-Derivatization Method: MPA

The determination of the absolute configuration of *sec/sec*-1,2-amino alcohols [62] (Figures 5.17 and 5.18a) follows a protocol similar to the one shown in Section 5.1 for *sec/sec*-1,2-diols [59–61]. It is based on the simultaneous derivatization of the

128 ■ The Assignment of the Absolute Configuration by NMR

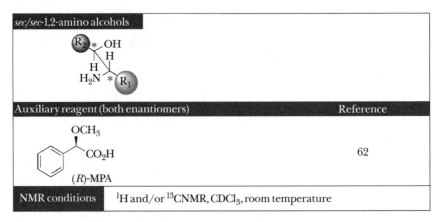

*Figure 5.17.* The assignment of *sec/sec*-1,2-amino alcohols by double-derivatization method at a glance: CDA, experimental conditions, and references.

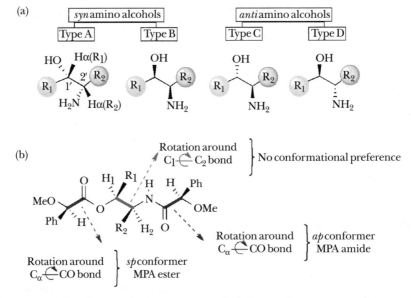

*Figure 5.18.* Classification of *sec/sec*-1,2-amino alcohols according to structural types and main bonds involved in conformational equilibria.

amino and alcohol groups, followed by the analysis of the chemical shifts of the resulting bis-(R)-MPA and bis-(S)-MPA amido esters derivatives [13, 62].

As in *sec/sec*-diols, the chemical shifts observed result from the combination of the shielding/deshielding effects transmitted by the two MPA units [62]. A correlation exists between the stereochemistry of the amino alcohols and the $\Delta\delta^{RS}$ signs of the diagnostic signals due to $H\alpha(R_1)/H\alpha(R_2)/R_1/R_2$ and $H\alpha(R_1)/H\alpha(R_2)$ in the *syn* and *anti*-1,2-amino alcohols, respectively (Figure 5.18a).

From the conformational point of view, the analysis of the shieldings in bis-MPA amido esters is similar to that presented for *sec/sec*-diols, with rotation around the Cα-CO bond in the MPA units being the only conformational process relevant for interpreting the NMR spectra.

Thus, the MPA ester moiety is preferentially in the *sp* conformation [37], whereas the MPA amide moiety is in the *ap* conformation [52] (Figure 5.18b), and this allows us to predict the shieldings and $\Delta\delta^{RS}$ signs for each isomer [62].

In a type A *syn*-1,2-amino alcohol, $H\alpha(R_1)$ and $R_1$ are located under the shielding cones in the bis-(R)-MPA derivative, and they are unaffected by the shielding cones in the bis-(S)-MPA derivative [62]. In the bis-(S)-MPA derivative, $H\alpha(R_2)$ and $R_2$ are under the shielding cones, and they are unaffected by them in the bis-(R)-MPA derivative [62]. Therefore, negative $\Delta\delta^{RS}$ signs are expected for $H\alpha(R_1)/R_1$ and positive ones for $H\alpha(R_2)/R_2$ (Figure 5.19a).

For the enantiomeric amino alcohol (type B), the opposite $\Delta\delta^{RS}$ sign distribution is predicted: positive $\Delta\delta^{RS}$ for $H\alpha(R_1)/R_1$ and negative for $H\alpha(R_2)/R_2$ (Figure 5.19b).

The bis-(R)- and bis-(S)-MPA derivatives of type C *anti*-1,2-amino alcohols (Figure 5.19c) and type D *anti*-1,2-amino alcohols (Figure 5.19d) have a notable characteristic: $R_1$ and $R_2$ are located under the shielding cones in both derivatives, and as a result, no correlation between the stereochemistry and the chemical shifts

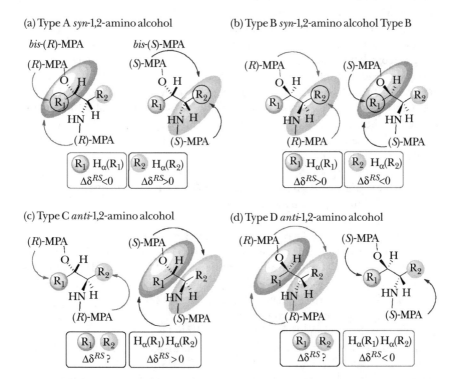

*Figure 5.19.* Shielding effects and $\Delta\delta^{RS}$ values for the bis-MPA esters of the four structural types of *sec/sec*-1,2-amino alcohols.

130 ■ The Assignment of the Absolute Configuration by NMR

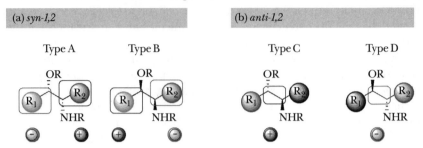

Figure 5.20. $\Delta\delta^{RS}$ sign distributions for acyclic *sec/sec*-1,2-amino alcohols (MPA as the CDA).

can be deduced. Therefore, the signals/shifts due to $R_1$ and $R_2$ are of no diagnostic value in the *anti*-amino alcohols [62]. In those cases, only $H\alpha(R_1)$ and $H\alpha(R_2)$ can be used, because we can establish a correlation between their behavior and their stereochemistry: in a type C *anti*-1,2-amino alcohol, $H\alpha(R_1)$ and $H\alpha(R_2)$ are under shielding cones in the bis-(S)-MPA ester, but this is not the case in the bis-(R)-MPA ester; thus a positive $\Delta\delta^{RS}$ is expected for both protons [62] (Figure 5.19c). The opposite trend is observed in a type D *anti*-1,2-amino alcohol, which has a negative $\Delta\delta^{RS}$ for $H\alpha(R_1)$ and $H\alpha(R_2)$ (Figure 5.19d).

A summary of $\Delta\delta^{RS}$ sign distributions for the bis-MPA derivatives of the four stereoisomers of a *sec/sec*-1,2-amino alcohol [62] is presented in Figure 5.20.

These correlations have been tested experimentally with a series of *sec/sec*-amino alcohols of known configuration, and there was complete agreement between the predictions from NMR and the experimentally proven absolute configurations [62] (see Figure 5.25 in Section 5.2.4). Finally, $^{13}$C-NMR chemical shifts can also be used for the assignment by treating the signs of $^{13}$C $\Delta\delta^{RS}$ in exactly the same way as we used those due to $^1$H-NMR. In any case, $^{13}$C $\Delta\delta^{RS}$ data should be complemented with $^1$H $\Delta\delta^{RS}$ data [72].

### 5.2.2. Example 33: Assignment of the Absolute Configuration of 2-Aminopentan-3-ol (*Syn*)

A sample of a stereoisomer of 2-aminopentan-3-ol (unknown configuration) is divided and separately derivatized with (R)- and (S)-MPA. The $^1$H-NMR spectra of the resulting bis-MPA amido esters are recorded (Figures 5.21a and b), the signals corresponding to $H\alpha(R_1)$, and $H\alpha(R_2)$ are assigned [i.e., H(2') and H(3')], and their $\Delta\delta^{RS}$ values are calculated (Figure 5.21c). In this way, opposite signs are obtained for those methine protons (+0.09 and −0.06 ppm, respectively), which indicates that the configuration of the amino alcohol is *syn*.

The signals from substituents $R_1$ and $R_2$ are assigned next, and their $\Delta\delta^{RS}$ values are calculated (Figure 5.21c). $R_1$ [i.e., H(4')/Me(5')] shows positive $\Delta\delta^{RS}$ values [+0.35/+0.44, respectively, the same sign as $H\alpha(R_1)$, +0.09 ppm], while $R_2$ [i.e., Me(1')] shows a negative $\Delta\delta^{RS}$ value [−0.46, the same sign as $H\alpha(R_2)$,

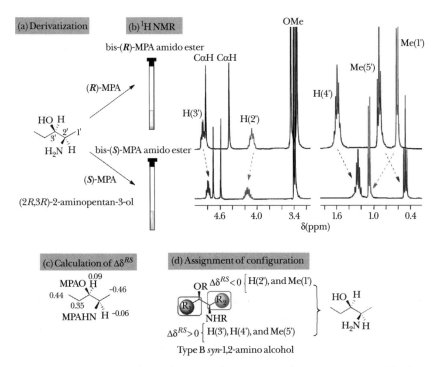

*Figure 5.21.* Main steps in the configurational assignment of 2-aminopentan-3-ol (*syn*) using MPA.

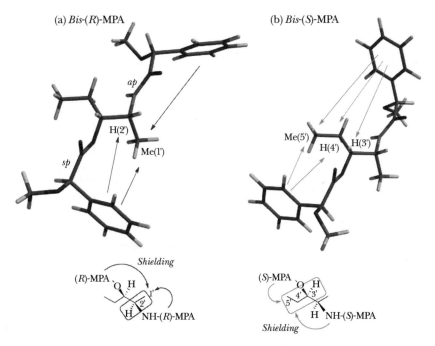

*Figure 5.22.* Position of the diagnostic hydrogens $H\alpha(R_1)/R_1$ and $H\alpha(R_2)/R_2$ with regard to the phenyl groups in the *sp* and *ap* conformers of the two bis-MPA amido esters of (2R, 3R)-2-aminopentan-3-ol.

−0.06 ppm]. Comparison of these $\Delta\delta^{RS}$ signs with those shown in Figure 5.20a shows that they match the type B *syn*-1,2-amino alcohol structure (Figure 5.21d) and correspond to the (2R, 3R) absolute configuration.

This stereochemistry and its relationship with the NMR spectra can be easily explained by the models shown in Figure 5.22, where the MPA ester moiety presents an *sp* conformation and the MPA amide moiety presents an *ap* conformation. Thus, Me(1′)/H(2′) are doubly shielded in the bis-(R)-MPA amido ester, while in the bis-(S)-MPA derivative they are not affected (Figure 5.22a), which leads to negative values of $\Delta\delta^{RS}$. On the other hand, H(3′)/H(4′)/Me(5′) are doubly shielded in the bis-(S)-MPA (Figure 5.22b) but not in the bis-(R)-MPA. As a consequence, their $\Delta\delta^{RS}$ values are positive.

### 5.2.3. Example 34: Assignment of the Absolute Configuration of Methyl 4-Amino-3-Hydroxy-5-Phenylpentanoate (*Anti*)

A sample of one isomer of methyl 4-amino-3-hydroxy-5-phenylpentanoate of unknown absolute configuration is divided and separately derivatized with (R)- and (S)-MPA, and the NMR spectra of the bis-MPA derivatives are registered

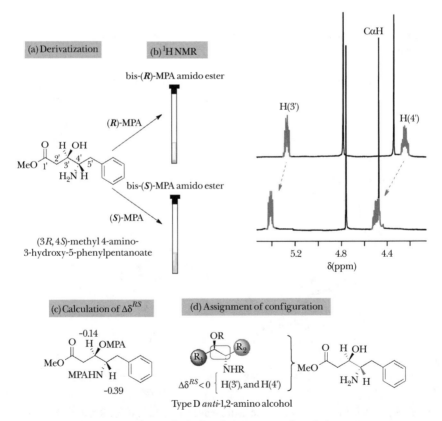

*Figure 5.23.* Main steps in the configurational assignment of methyl 4-amino-3-hydroxy-5-phenylpentanoate (*anti*) using MPA.

(Figures 5.23a and b). The first step is to determine if the configuration of the amino alcohol is *syn* or *anti* by examination of the methine protons [i.e., $H\alpha(R_1)$ and $H\alpha(R_2)$]. We find that H(3′) and H(4′) have negative $\Delta\delta^{RS}$ signs (i.e., −0.14 and −0.39 ppm, respectively), which indicates that the amino alcohol is *anti*. Consequently, it is not necessary to assign the $R_1$ and $R_2$ signals in the spectra. As seen in Figure 5.20b, negative $\Delta\delta^{RS}$ values for $H\alpha(R_1)$ and $H\alpha(R_2)$ are characteristic of type D *anti*-1,2-amino alcohols.

Therefore, the amino alcohol under scrutiny is a type D *anti*-1,2-amino alcohol with the (3R, 4S) configuration (Figure 5.23d).

Those shifts can be explained using the models shown in Figure 5.24. In the bis-(R)-MPA amido ester (Figure 5.24a), the phenyl rings of both MPA moieties transmit their shielding effects to H(2′)/H(3′)/H(4′)/H(5′). In the bis-(S)-MPA derivative (Figure 5.24b), protons H(2′)/H(5′) are shielded, and thus those signals are of no diagnostic value; the H(3′)/H(4′) are unaffected. As a result, this combination of anisotropic effects leads to negative $\Delta\delta^{RS}$ values for H(3′)/H(4′), which are the diagnostic protons.

### 5.2.4. Summary

The absolute configuration of the two asymmetric carbons found in a *sec/sec*-amino alcohol can be simultaneously determined by double derivatization [13, 62] with (R)- and (S)-MPA and comparison of the ¹H NMR of the bis-(R)- and the bis-(S)-MPA amido ester derivatives.

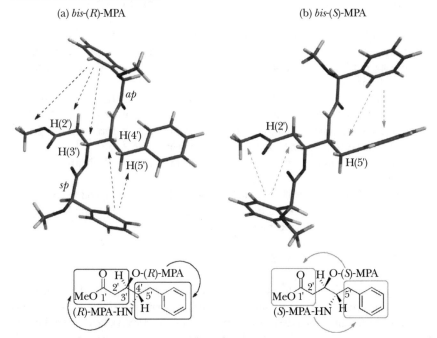

*Figure 5.24.* Position of the diagnostic hydrogens $H\alpha(R_1)$ and $H\alpha(R_2)$ with regard to the phenyl groups in the *sp* and *ap* conformers of the two bis-MPA amido esters of (3R, 4S)-methyl 4-amino-3-hydroxy-5-phenylpentanoate.

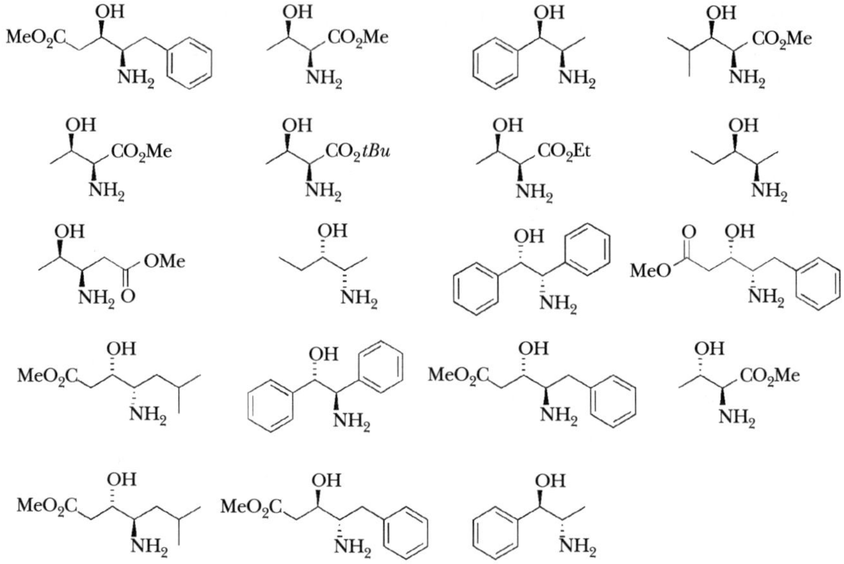

*Figure 5.25.* Selection of *sec/sec*-1,2-amino alcohols with known absolute configuration that have been used to validate the procedures described in this chapter.

As in *sec/sec*-diols, the two functional groups are derivatized simultaneously [62]. The methine protons directly linked to the asymmetric carbons [i.e., $H\alpha(R_1)$ and $H\alpha(R_2)$] are the signals used for diagnosis. Each of the four possible stereoisomers present a specific distribution of $\Delta\delta^{RS}$ signs for those protons.

In the case of *syn*-1,2-amino alcohols, once the sign pattern of the methines identifies a particular stereoisomer, the $\Delta\delta^{RS}$ signs for $R_1$ and $R_2$ can be used to confirm the assignment.

The exclusive use of $^{13}$C NMR for the assignment of the configuration of *sec/sec*-1,2-amino alcohols is not recommended [72], because the $^{13}$C $\Delta\delta^{RS}$ values are frequently too small to ensure a safe assignment. Nevertheless, if significant $^{13}$C $\Delta\delta^{RS}$ data is available, it can be used to complement $^1$H $\Delta\delta^{RS}$ data.

To date, no method based on the NMR spectra of a single derivative has been described.

Figure 5.25 shows a selection of amino alcohols of known absolute configuration used to validate the procedures of this chapter [62]. Other examples of applications of this methodology can be found in the literature [214].

## 5.3. PRIM/SEC-1,2-DIOLS

The assignment of the absolute configuration of compounds formed by a primary hydroxyl group and a vicinal secondary (chiral) hydroxyl group (*prim/sec*-1,2-diols) involves the simultaneous derivatization of the two hydroxyls, leading to the corresponding bis-(*R*)- and bis-(*S*)-CDA esters and the comparison of their $^1$H-NMR spectra [13, 63–65] (Figure 5.26). MPA and 9-AMA are the CDAs of choice. If

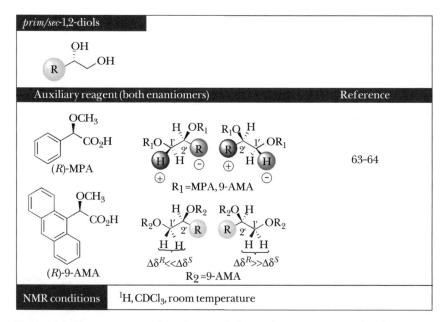

*Figure 5.26.* The assignment of *prim/sec*-1,2-diols at a glance: CDAs, sign distribution ($\Delta\delta^{RS}$), methylene hydrogens δ differences ($\Delta\delta^R$, $\Delta\delta^S$), experimental conditions and references.

MPA is employed, the assignment of the configuration is carried out by double derivatization followed by comparison of the $\Delta\delta^{RS}$ signs from the R and the methylene [63] (1′) substituents. If 9-AMA is chosen instead of MPA, the assignment can be performed in two different ways [64]: (a) through the $\Delta\delta^{RS}$ signs from the protons of R and the methylene (1′) groups (likewise for the use of MPA), or (b) in a simplified way, just by analysis of the $^1$H-NMR chemical shifts of the diastereotopic methylene protons (1′).

### 5.3.1. Double-Derivatization Methods: MPA

As indicated above (Section 5.3), the absolute configuration of *prim/sec*-1,2-diols can be determined by double derivatization [13, 63] with MPA in a way similar to that described for secondary alcohols (Section 3.1.1) [14, 15, 35–39]. In this way, one stereoisomer shows a certain $\Delta\delta^{RS}$ sign for the R protons and the opposite sign for the methylene (1′) protons, and this sign distribution is reversed in the enantiomeric diol structure [63]. The correlation between the $\Delta\delta^{RS}$ sign distribution and the stereochemistry for *prim/sec*-1,2-diols derivatized with two MPA units fits almost exactly with the one observed for mono-MPA derivatives of secondary alcohols with the same configuration [37]. The only discrepancy appears to be that one of the methylene protons usually shows either a very small $\Delta\delta^{RS}$ value or a sign opposite to the expected one (i.e., of no diagnostic value). This anomaly is due to the MPA unit at the primary hydroxyl, which modifies the conformation around the O–C(1′) bond in such a way that the combined anisotropic effects of the closest carbonyl and the two phenyls affect the two methylene hydrogens in different ways. However, the $\Delta\delta^{RS}$ sign distributions from the R group and the other methylene hydrogen allow the reliable assignment of the absolute configuration of these systems.

Conformational analysis [63] of bis-MPA esters of *prim/sec*-1,2-diols, based on theoretical calculations and experimental data (Figure 5.27), shows that the most important processes are those related to the rotations around the Cα-C(O) bonds (*sp* and *ap* conformers), the C(1′)–C(2′) bond (*gauche–trans* conformer; *gt*), and the O–C(1′) bond (*I* and *II* conformers; Figure 5.27). Of all of these, the NMR spectra of *sp-gt-I* is the most representative for the conformation of bis-(*R*)-MPA esters, while bis-(*S*)-MPA esters are best represented by the *sp-gt-II* form (see Figure 5.27).

Considering those main conformations [63], it is possible to predict the shielding/deshielding effects generated in each derivative. Thus, the R group in the bis-(*R*)-MPA ester, with the absolute configuration at C(2′) [type A] shown in Figure 5.28a, is shielded by the phenyls of the two MPA units, whereas in the bis-(*S*)-MPA ester, it is shielded only by the MPA unit that esterifies the primary

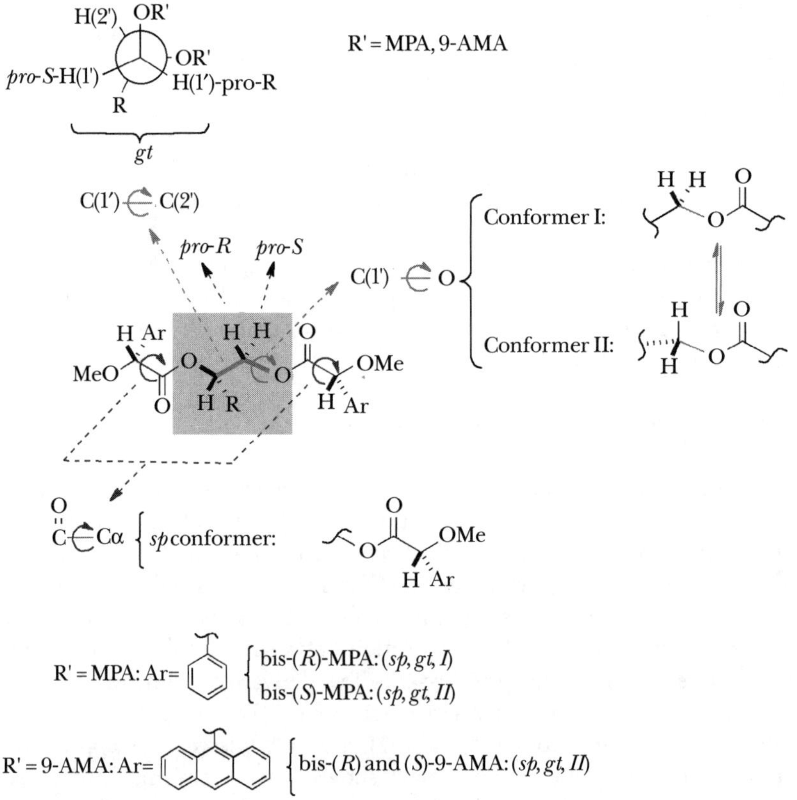

*Figure 5.27.* Main bonds and conformers involved in the conformational equilibria of *prim/sec*-1,2-diols.

alcohol (Figure 5.28b). The combined result of these effects is that the R group produces a negative $\Delta\delta^{RS}$ sign for that stereochemistry (Figure 5.28c).

As indicated above, the shielding/deshielding effects on the methylene protons are more complex [63, 65]. In the bis-(R)-MPA ester of the same stereoisomer, the conformation around the O–C(1′) bond is of type I (Figures 5.27 and 5.28a), and the carbonyl bisects the angle between the two methylene protons (each one separated by about 60° from the carbonyl plane); this affects them both in the same way.

On the other hand, in the bis-(S)-MPA ester, the conformation around the O–C(1′) bond is of type II, which means that the carbonyl is coplanar with the pro-S-H(1′) hydrogen (Figure 5.27 and 5.28b). As a consequence of this situation, the pro-R-H(1′) methylene hydrogen is triple shielded by the phenyls of both MPA units and by the carbonyl group of the MPA that esterifies the primary alcohol (Figure 5.28b). Thus, pro-R-H(1′) has a large and positive $\Delta\delta^{RS}$ value

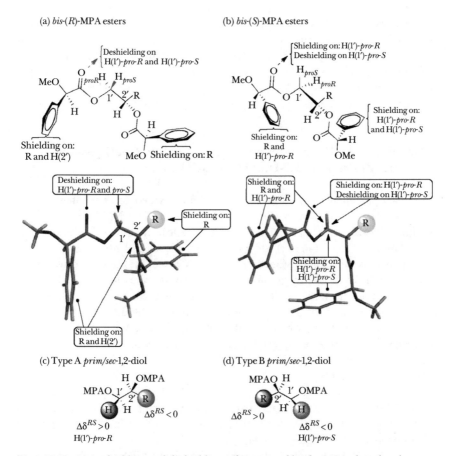

*Figure 5.28.* Main shielding and deshielding effects caused by the MPA phenyl and carbonyl groups on the R and methylene (1′) hydrogens in bis-(R)-MPA and bis-(S)-MPA esters of *prim/sec*-1,2-diols.

138 ■ The Assignment of the Absolute Configuration by NMR

(Figure 5.28c). However, the pro-S-H(1') methylene hydrogen is shielded by the phenyl of the MPA that esterifies the secondary alcohol, and it is deshielded by the carbonyl of the MPA that esterifies the primary alcohol (Figure 5.28b). Therefore, the pro-S-H(1') has a very small $\Delta\delta^{RS}$ value with a sign that is difficult to predict reliably. This prevents the stereochemical assignment of the methylene hydrogen with the smallest $\Delta\delta^{RS}$ value, and leaves the methylene hydrogen with the largest $\Delta\delta^{RS}$ value as the only one with diagnostic value [63].

If the configuration at C(2') is the opposite form (type B), the $\Delta\delta^{RS}$ values are those shown in Figure 5.28d. It is necessary to point out that in this case, and as a result of a similar analysis as above, pro-S-H(1') is the methylene hydrogen that shows the largest value, and it is acceptable for assignment.

These correlations between $\Delta\delta^{RS}$ signs and stereochemistry have been experimentally demonstrated with a series of compounds of known absolute configuration [63–65]. A selection is shown in Figure 5.46 in Section 5.3.10.

To sum up, the absolute configuration of *prim/sec*-1,2-diols can be assigned by double derivatization [63] in the following way: the two bis-MPA esters are prepared, their $^1$H-NMR spectra are examined, and the chemical shift differences (i.e., $\Delta\delta^{RS}$) from the hydrogens of the R group and from the methylene (1') hydrogen that shows the largest $\Delta\delta^{RS}$ value, are calculated. If the $\Delta\delta^{RS}$ of R is negative and that of the methylene (1') hydrogen is positive, the configuration of the diol is that shown in Figure 5.28c (type A). If they have opposite signs [i.e., $\Delta\delta^{RS}$ is positive for R and negative for the methylene (1') hydrogen], the configuration of the diol is that shown in Figure 5.28d (type B).

### 5.3.2. Example 35: Assignment of the Absolute Configuration of (*S*)-Propane-1,2-Diol Using MPA

A sample of an enantiomer of propane-1,2-diol is divided and separately derivatized with (*R*)- and (*S*)-MPA (Figure 5.29a). Next, the $^1$H-NMR spectra of the resulting bis-MPA esters are registered (Figure 5.29b), and the $\Delta\delta^{RS}$ values are calculated for Me(3') (–0.14 ppm) and for both methylene (1') hydrogens (+0.22 and +0.03 ppm). As shown above (Section 5.3.1), only one of the methylene hydrogens (the one with the largest magnitude $\Delta\delta^{RS}$) is used as a diagnostic signal. In this case, the diagnostic value is +0.22 ppm. The absolute configuration is assigned using the distributions of $\Delta\delta^{RS}$ signs shown in Figures 5.28c and d.

The negative and positive signs obtained from R [i.e., Me(3')] and from the methylene hydrogen, respectively, indicate that the absolute configuration corresponds to the one shown in Figure 5.28c (type A). Thus, the diol is (*S*)-propane-1,2-diol (Figure 5.29d).

The different chemical shifts observed in the $^1$H-NMR spectra shown in Figure 5.29b can be easily explained and justified by considering the *sp-gt-I* conformation for the bis-(*R*)-MPA ester and the *sp-gt-II* for the bis-(*S*)-MPA ester of (*S*)-propane-1,2-diol (Figures 5.28 and 5.30).

Me(3') is shielded by both MPA esters in the bis-(*R*)-MPA ester (Figure 5.30a), while in the bis-(*S*)-MPA ester, it is not affected (Figure 5.30b). As this group is more shielded in the bis-(*R*)- than in the bis-(*S*)-MPA ester, it presents a negative $\Delta\delta^{RS}$ value.

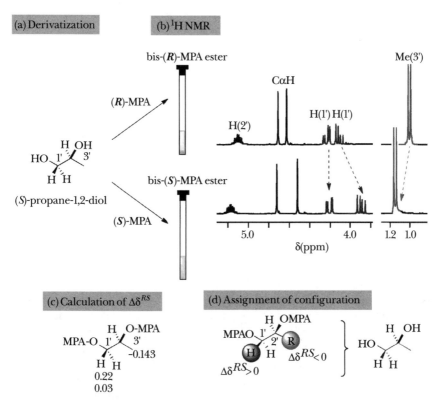

Figure 5.29. Main steps in the configurational assignment of (S)-propane-1,2-diol using MPA.

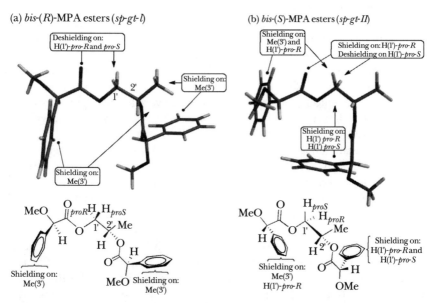

Figure 5.30. Main shielding and deshielding effects caused by the MPA phenyl and carbonyl groups on the R and methylene (1′) hydrogens in the bis-(R)-MPA and bis-(S)-MPA esters of (S)-propane-1,2-diol.

On the other hand, in the bis-(R)-MPA ester (Figure 5.30a), both methylene (1′) hydrogens are affected in the same way by the carbonyl group of the MPA at the primary hydroxyl. In the bis-(S)-MPA ester (Figure 5.30b), one of the methylene hydrogens is shielded by the phenyls of the two MPAs and the carbonyl of the MPA at the primary hydroxyl, and therefore it presents a positive $\Delta\delta^{RS}$ value. The other methylene proton is shielded by the phenyl of the MPA at the secondary hydroxyl and deshielded by the carbonyl of the MPA at the primary hydroxyl; thus its $\Delta\delta^{RS}$ is small, and it is difficult to predict its sign.

### 5.3.3. Double-Derivatization Methods: 9-AMA

The configuration of *prim/sec*-1,2-diols can also be determined through the bis-9-AMA esters [13, 64]. In this case, conformational analysis shows that the most representative conformers are the same in both bis-9-AMA-esters [37]. This is summarized as follows: the two 9-AMA units are in *sp* conformations, the preference around the C(1′)–C(2′) bond is *gt* and around the O–C(1′) bond, it is a type II conformation [13] (Figure 5.27). The ¹H-NMR spectra can therefore be easily predicted by considering these conformations. For the model compounds shown in Figures 5.31a, b, and c (type A), the R group is more shielded in the bis-(R)-9-AMA (Figure 5.31a) than in the bis-(S)-9-AMA (Figure 5.31b), and as a consequence its $\Delta\delta^{RS}$ is negative (Figure 5.31c).

In relation to the behavior of the methylene (1′) hydrogens, a more detailed analysis needs to be performed. Pro-R-H(1′) is shielded by the carbonyl of the 9-AMA that esterifies the primary alcohol in the bis-(R)-9-AMA ester (Figure 5.31a). In the bis-(S)-9-AMA, it is shielded by the carbonyl and by the anthryls of each of the 9-AMA units (Figure 5.31b). Therefore, it shows a large and positive $\Delta\delta^{RS}$ value (Figure 5.31c). Pro-S-H(1′) is deshielded by the carbonyl of the 9-AMA that esterifies the primary alcohol in the bis-(R)-9-AMA, whereas in the bis-(S)-9-AMA, it is deshielded by the carbonyl and shielded by the anthryl of the 9-AMA that esterifies the secondary alcohol. Therefore, its $\Delta\delta^{RS}$ is small, and the sign is difficult to predict. This eliminates this methylene hydrogen [i.e., pro-S-H(1′)] from being used for stereochemical diagnosis. The methylene hydrogen of diagnostic value is the one with the larger $\Delta\delta^{RS}$ value [i.e., pro-R-H(1′)], as was also the case with the bis-MPA-derivatives.

If the configuration at C(2′) is the opposite form (type B), the $\Delta\delta^{RS}$ values are those shown in Figure 5.31d. In this case, pro-S-H(1′) is the methylene hydrogen that shows the largest value and is thus valid for diagnosis [13].

Therefore, the absolute configuration of *prim/sec*-1,2-diols can be assigned using the double-derivatization method [13]: the two bis-9-AMA esters are prepared, and the $\Delta\delta^{RS}$ signs are determined for the hydrogens of the R and the methylene (1′) groups. In the latter case, only the hydrogen that shows the largest $\Delta\delta^{RS}$ magnitude is used as a diagnostic sign. If the $\Delta\delta^{RS}$ of R is negative and that of the methylene (1′) is positive, the configuration of the diol is that shown in Figure 5.31c (type A). If, on the contrary, $\Delta\delta^{RS}$ is positive for R and negative for the methylene (1′) hydrogen, the diol has the configuration shown in Figure 5.31d (type B).

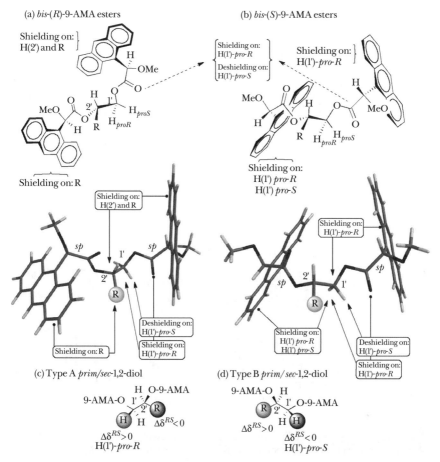

*Figure 5.31.* Main shielding and deshielding effects caused by the 9-AMA anthryl and carbonyl groups on the R and methylene (1′) hydrogens in bis-(R)-9-AMA and bis-(S)-9-AMA esters of *prim/sec*-1,2-diols.

The NMR pattern of the methylene (1′) hydrogens in the bis-(R)- and the bis-(S)-derivatives follows a trend that correlates with the absolute stereochemistry of the *prim/sec*-1,2-diols and allows the formulation of a simplified procedure for assignment [64, 165]. This method requires the preparation of the two bis-9-AMA esters, and it requires the calculation of the $\Delta\delta^{RS}$ values only for the methylene protons.

The models of the two derivatives in the most representative conformers (i.e., *sp-gt-II*) of an enantiomer of a *prim/sec*-1,2-diol (Figures 5.32a and b; type A) show how the shielding effects and chemical shifts of the methylene (1′) hydrogens correlate with the configuration of that diol.

In the bis-(R)-9-AMA, the two methylene hydrogens are affected in opposite ways by the anisotropy of the carbonyl of the 9-AMA ester at the primary alcohol, but they are not affected by any anthryl group (Figure 5.32a). In this situation,

142 ■ The Assignment of the Absolute Configuration by NMR

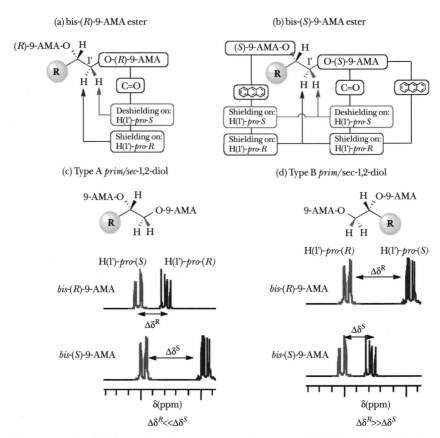

*Figure 5.32.* Simplified procedure for assignment of *prim/sec*-1,2-diols based on $\Delta\delta^R$ and $\Delta\delta^S$ parameters from the diastereotopic methylene hydrogens in bis-9-AMA esters.

pro-S-H(1′) is slightly deshielded, and pro-R-H(1′) is slightly shielded [13, 64]. The overall result is that the two protons resonate at relatively similar frequencies, because the carbonyl effects are not as intense as are the aromatic ones (Figure 5.32c).

In the bis-(S)-9-AMA ester of the same diol, pro-S-H(1′) is deshielded by the carbonyl of the 9-AMA unit at the primary alcohol and shielded by the anthryl of the 9-AMA unit that esterifies the secondary hydroxyl (Figure 5.32b). Pro-R-H(1′) experiences a triple shielding by the carbonyl and by the two anthryls (Figure 5.32b). As a result of such different shieldings on the two methylene protons, they resonate at very different frequencies [13, 64] (Figure 5.32c).

In summary, a *prim/sec*-1,2-diol with the configuration shown in Figure 5.32c presents the methylene (1′) protons resonating at closer frequencies in the bis-(R)-9-AMA than in the bis-(S)-9-AMA esters, that is to say, the separation between the methylene (1′) protons is greater in the bis-(S)-9-AMA derivative than in the bis-(R)-9-AMA derivative (i.e., $\Delta\delta^R \ll \Delta\delta^S$, where $\Delta\delta^R = \delta H(1′)$

(downfield) − δH(1′)(upfield) in the bis-(R)-9-AMA derivative and $\Delta\delta^S$ = δH(1′) (downfield) − δH(1′)(upfield) in the bis-(S)-9-AMA derivative).

The enantiomeric *prim/sec*-1,2-diol (Figure 5.32d; type B) shows the reverse distribution of shielding effects: the methylene protons resonate at closer frequencies in the bis-(S)-9-AMA than in the bis-(R)-9-AMA esters (i.e., $\Delta\delta^R \gg \Delta\delta^S$).

### 5.3.4. Example 36: Assignment of the Absolute Configuration of (*S*)-Propane-1,2-Diol Using 9-AMA

To carry out the assignment of the absolute configuration of an enantiomer of propane-1,2-diol, two small amounts of this diol are separately derivatized with (R)- and (S)-9-AMA (Figure 5.33a). The $^1$H-NMR spectra of the resulting bis-MPA esters are registered (Figure 5.33b), and the $\Delta\delta^{RS}$ parameters are calculated for Me(3′) (−0.15 ppm) and for both methylene (1′) protons (+0.21 and +0.03 ppm). The absolute configuration is assigned by comparison with the distribution of the $\Delta\delta^{RS}$ signs shown in Figures 5.31c and d. The negative sign from R [i.e., Me(3′); −0.15 ppm] and the positive one from the diagnostic methylene (1′) hydrogen (i.e., the one with the largest magnitude; +0.21 ppm)

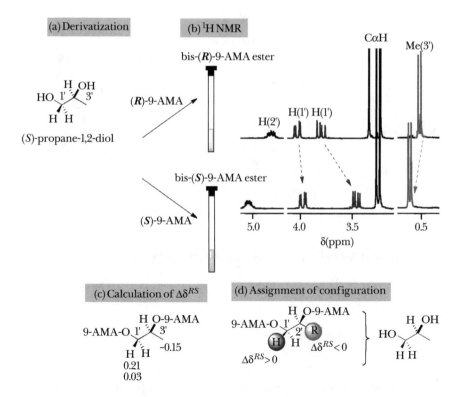

*Figure 5.33.* Main steps in the configurational assignment of (*S*)-propane-1,2-diol using 9-AMA.

### 5.3.5. Example 37: Assignment of the Absolute Configuration of (R)-2,3-Dihydroxypropyl Stearate Based Only on the Methylene Hydrogens

A sample of 2,3-dihydroxypropyl stearate of unknown absolute configuration is divided and separately derivatized with (R)- and (S)-9-AMA (Figure 5.34a). The $^1$H-NMR spectra of the resulting bis-9-AMA esters are registered, and the spectral region corresponding to the methylene (1′) hydrogens is analyzed (Figure 5.34b). The separations between the signals of those two hydrogens [i.e., $CH_2(1′)$] are calculated for the bis-(R)-9-AMA (i.e., $\Delta\delta^R$) and the bis-(S)-9-AMA (i.e., $\Delta\delta^S$) derivatives. The following values are obtained: $\Delta\delta^R$ = +0.16 ppm and $\Delta\delta^S$ = +0.47 ppm (Figure 5.34c). As the difference is larger in the bis-(S)-9-AMA ester than in the bis-(R)-9-AMA ester (i.e., $\Delta\delta^R \ll \Delta\delta^S$), the diol has the configuration shown in Figure 5.34d [i.e., (R)-2,3-dihydroxypropyl stearate].

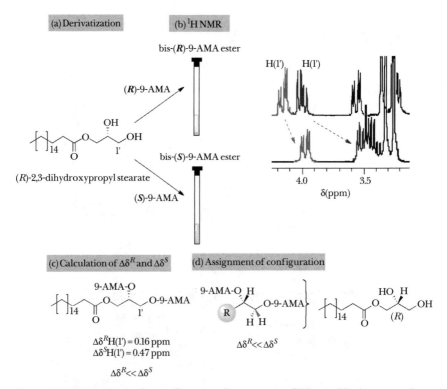

*Figure 5.34.* Main steps in the configurational assignment of (R)-2,3-dihydroxypropyl stearate based on the methylene protons using 9-AMA. Note: the (1′) numbering is kept for the methylene group attached to the primary alcohol to maintain coherence with the numbering of the models.

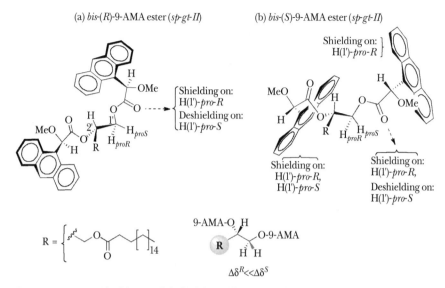

*Figure 5.35.* Main shielding and deshielding effects caused by the 9-AMA anthryl and carbonyl groups on methylene (1′) hydrogens in bis-(R)-9-AMA and bis-(S)-9-AMA esters of (R)-2,3-dihydroxypropyl stearate.

These $^1$H-NMR spectra patterns can be easily explained by considering the main *sp-gt-II* conformations [13, 64] of the bis-9-AMA esters (Figure 5.35). In the bis-(R)-9-AMA ester, pro-S-H(1′) is deshielded by the carbonyl of the 9-AMA that esterifies the primary alcohol, while the pro-R-H(1′) is shielded (Figure 5.35a). Therefore, the two hydrogens resonate at relatively similar frequencies (i.e., small $\Delta\delta^R$).

In the bis-(S)-9-AMA ester, the pro-S-H(1′) is deshielded by the carbonyl of the 9-AMA ester that esterifies the primary alcohol and shielded by the anthryl of the 9-AMA that esterifies the secondary alcohol (Figure 5.35b). On the other hand, pro-R-H(1′) experiences a triple shielding: by the carbonyl and by the two anthryls (Figure 5.35b). Therefore, the two methylene (1′) hydrogens of the bis-(S)-9-AMA ester resonate at very different frequencies (i.e., large $\Delta\delta^S$).

In summary, the separation between the signals of the two methylene hydrogens is larger in the bis-(S)-9-AMA ester than in the bis-(R)-9-AMA one ($\Delta\delta^R \ll \Delta\delta^S$).

### 5.3.6. Example 38: Assignment of the Absolute Configuration of (R)-1-Phenylethane-1,2-Diol Based Only on the Methylene Hydrogens

A sample of 1-phenylethane-1,2-diol of unknown absolute configuration is divided and separately derivatized with (R)- and (S)-9-AMA (Figure 5.36a). The $^1$H-NMR spectra of the resulting bis-9-AMA esters are registered, and the zone corresponding to the resonances of the methylene (1′) hydrogens is analyzed (Figure 5.36b).

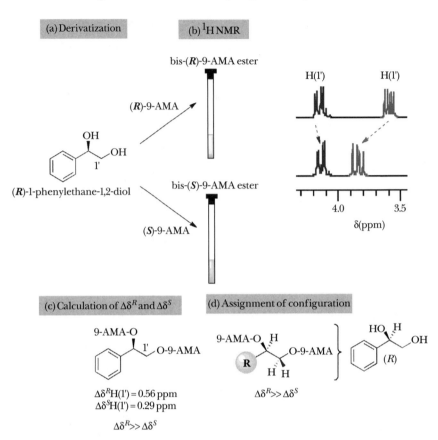

*Figure 5.36.* Main steps in the configurational assignment of (R)-1-phenylethane-1,2-diol based on the methylene protons using 9-AMA. Note: the (1′) numbering is kept for the methylene group attached to the primary alcohol to maintain coherence with the numbering of the models.

The separations between the signals of those hydrogens are calculated and found to be $\Delta\delta^R$= +0.56 ppm and $\Delta\delta^S$= +0.29 ppm. As the separation between the methylene signals is greater in the bis-(R)- than in the bis-(S)-9-AMA derivative (i.e., $\Delta\delta^R \gg \Delta\delta^S$), the configuration of the diol corresponds to that shown in Figure 5.32d [(R)-1-phenylethane-1,2-diol].

Again, the shieldings and spectra can be interpreted on the basis of Dreiding models representing the main *sp-gt-II* conformations [13, 64] of the bis-9-AMA esters (Figure 5.37). In the bis-(R)-9-AMA ester, the pro-S-H(1′) experiences a triple shielding by the carbonyl of the 9-AMA that esterifies the primary alcohol and the two anthryls (Figure 5.37a), and the pro-R-H(1′) is deshielded by the carbonyl of the same 9-AMA and shielded by the anthryl of the 9-AMA that esterifies the secondary hydroxyl (Figure 5.37a). Therefore, the two methylene protons resonate at very different frequencies (i.e., large $\Delta\delta^R$).

In the bis-(S)-9-AMA ester, the pro-S-H(1′) is shielded by the carbonyl of the 9-AMA that esterifies the primary alcohol, while the pro-R-(1′) is deshielded

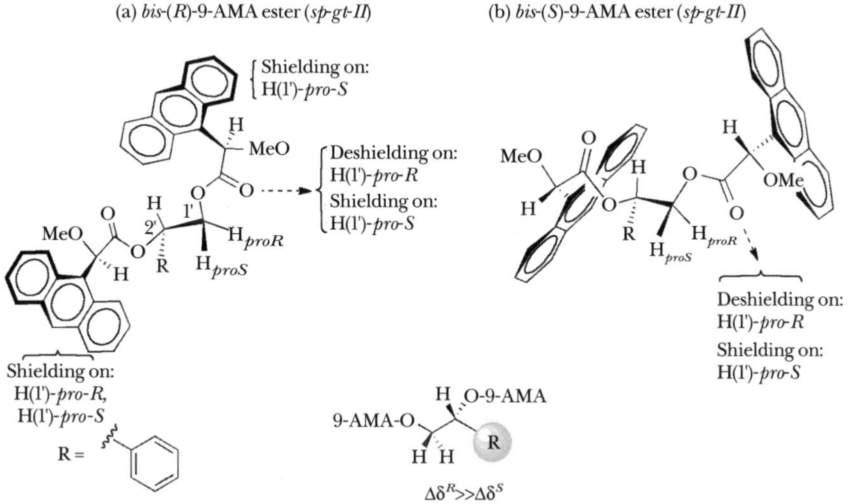

*Figure 5.37.* Main shielding and deshielding effects caused by the 9-AMA anthryl and carbonyl groups on methylene (1′) hydrogens in bis-(R)-9-AMA and bis-(S)-9-AMA esters of (R)-1-phenylethane-1,2-diol.

(Figure 5.37b). Therefore, the two protons resonate at relatively similar frequencies (i.e., small $\Delta\delta^S$). As a result, the separation between the methylene signals is larger in the bis-(R)-9-AMA derivative than in the bis-(S)-9-AMA one (i.e., $\Delta\delta^R \gg \Delta\delta^S$).

## 5.3.7. Single-Derivatization Method: MPA

The assignment of absolute configuration by comparison of the $^1$H-NMR spectra at two temperatures was amply discussed for monofunctional compounds [41, 165] (Chapter 4), and it is based on: (1) the existence of a conformational equilibrium with a well-defined major conformer that is similar in the (R)-CDA and (S)-CDA derivatives, and (2) the presence in the substrate of substituents on either side of the plane defined by the asymmetric carbon, and that are differently affected by the aromatic anisotropy generated by the CDA in the major conformer. In this way, a decrease in the probe temperature is accompanied by an increase in the contribution of the major conformer to the average NMR spectrum and a decrease in the second more abundant conformer [36, 41]. These changes in the population of the conformers in equilibrium generate the corresponding shifts of the signals of the substituents around the chiral center to a higher field (positive $\Delta\delta^{T1,T2}$) or to a lower field (negative $\Delta\delta^{T1,T2}$) [41, 165]. Typical examples can be found in Chapter 4.

In the case of polyfunctional compounds such as *sec/sec*-1,n-diols, which have two asymmetric carbons and two pairs of stereoisomers, this method allows for the distinction and assignment of three out of the four possible isomers [61]. That is to say, it allows the assignment of the absolute configuration of each

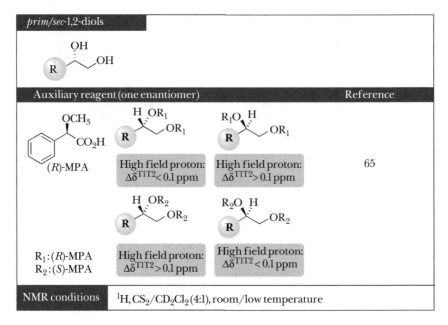

*Figure 5.38.* The assignment of *prim/sec*-1,2-diols by the single-derivatization method at a glance: CDA, diagnostic methylene hydrogen shifts ($\Delta\delta^{T1T2}$), experimental conditions, and references.

of the two *anti*-1,n-diols but only the relative stereochemistry for the *syn* pair (Section 5.1.5).

In this section (Figure 5.38), we will show that the absolute configuration of *prim/sec*-1,2-diols can be effectively determined by the low-temperature procedure with the bis-MPA ester derivatives using the methylene (1') hydrogens as diagnostic signals [65, 165] (Figure 5.39).

Thus, in the bis-(*R*)-MPA ester of a *prim/sec*-1,2-diol with the configuration shown in Figure 5.39a (type A), the conformational equilibrium [63] is established between the conformers *sp-gt-I* and *ap-gt-I* (Section 5.3.1). In both conformers, the main anysotropic effect affecting the methylene (1') hydrogens is the one from the carbonyl of the MPA unit at the primary alcohol; the pro-*R*-H(1') is slightly shielded and the pro-*S*-H(1') is slightly deshielded (Figure 5.39a).

When the NMR probe temperature decreases, the most stable conformer increases its contribution to the average spectrum [41, 65], and the hydrogen that resonates at the highest field [i.e., pro-*R*-H(1')] is slightly more shielded (less than 0.1 ppm), whereas the hydrogen that resonates at the lowest field [i.e., pro-*S*-H(1')] is slightly deshielded (Figure 5.39b). Therefore, the $\Delta\delta^{T1T2}$ is positive for the pro-*R*-H(1') (i.e., $\Delta\delta^{T1T2} > 0$) and negative for the pro-*S*-H(1') (i.e., $\Delta\delta^{T1T2} < 0$ (Figure 5.39b).

For the enantiomeric *prim/sec*-1,2-diol (type B, Figure 5.40a), the conformational equilibrium of its bis-(*R*)-MPA ester is established between the *sp-gt-II* and

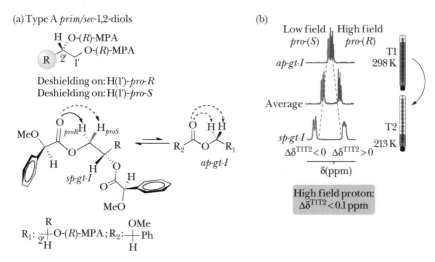

Figure 5.39. Simplified procedure for assignment of type A *prim/sec*-1,2-diols based on $\Delta\delta^{T1T2}$ parameters from the diastereotopic methylene hydrogens in bis-(R)-MPA esters.

*ap-gt-I* conformers, the former being the most stable. In the *sp-gt-II* conformer, the hydrogen that resonates at the highest field [i.e., pro-S-H(1′)] is triple shielded by the phenyls of the two MPA units and by the carbonyl of the MPA at the primary alcohol. At a lower temperature, the population of the most stable conformer increases, leading to a strong upfield shift (i.e., $\Delta\delta^{T1T2} > 0$), which is typically between 0.2 and 0.3 ppm (Figure 5.40b).

In the most stable conformation, the hydrogen that resonates at the lowest field [i.e., pro-R-H(1′)] is shielded by a phenyl and deshielded by the carbonyl. As the temperature decreases, this proton shifts slightly downfield (i.e., $\Delta\delta^{T1T2} < 0$; 0.0–0.1 ppm; Figure 5.40b).

The same diols esterified with (S)-MPA show the opposite behavior with temperature [13, 65, 165].

In practice, to assign the configuration of a *prim/sec*-diol by using low-temperature NMR of a single bis-MPA-ester, it is only necessary to analyze the NMR behavior of the methylene proton that resonates at the higher field. This method requires the following steps [65]: (a) Preparation of either the bis-(R)- or the bis-(S)-MPA derivative; (b) comparison of its $^1$H-NMR spectra taken at room temperature (around 300 K) and at a lower temperature (typically between 180 and 240 K); (c) calculation of the $\Delta\delta^{T1T2}$ values for the methylene proton that resonates at the higher field; and (d) if it is a bis-(R)-MPA ester and $\Delta\delta^{T1T2}$ for the high-field proton is < 0.10 ppm, then the diol has the configuration shown in Figure 5.41a (type A); if $\Delta\delta^{T1T2} > 0.10$ ppm, then the absolute configuration of the diol is shown in Figure 5.41b (type B).

For the bis-(S)-MPA ester, the opposite holds: if $\Delta\delta^{T1T2} > 0.10$ ppm, the diol has the configuration shown in Figure 5.41c (type A), and if $\Delta\delta^{T1T2} < 0.10$ ppm, it has the configuration shown in Figure 5.41d (type B).

150 ■ The Assignment of the Absolute Configuration

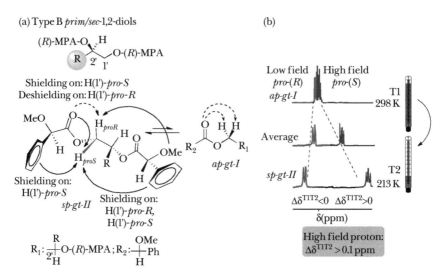

*Figure 5.40.* Simplified procedure for assignment of type B *prim/sec*-1,2-diols based on $\Delta\delta^{T1T2}$ parameters from the diastereotopic methylene hydrogens in bis-(*R*)-MPA esters.

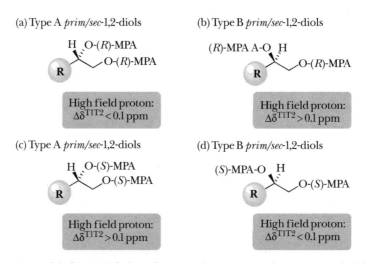

*Figure 5.41.* Models for simplified configurational assignment of *prim/sec*-1,2-diols based on $\Delta\delta^{T1T2}$ parameters from the high-field methylene (1′) hydrogen in bis-MPA esters.

### 5.3.8. Example 39: Assignment of the Absolute Configuration of (*S*)-Propane-1,2-Diol

Figure 5.42 shows schematically the steps (a–f) for the assignment of the absolute configuration of a sample of an enantiomer of propane-1,2-diol derivatized with (*R*)-MPA. Four ¹H-NMR spectra are registered from 300 K to 183 K, and the

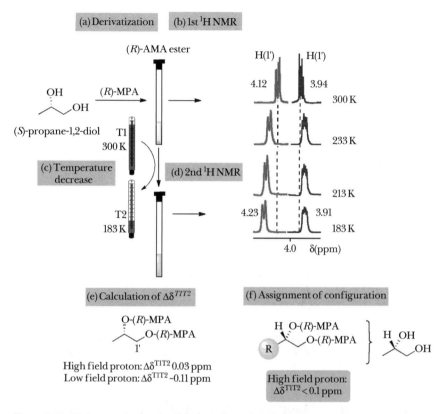

*Figure 5.42.* Main steps in the simplified configurational assignment of (S)-propane-1,2-diol based on the high-field methylene (1′) hydrogen using (R)-MPA.

$\Delta\delta^{T1T2}$ parameter that is calculated for the high-field methylene (1′) proton is 0.03 ppm. As this value is <0.10 ppm, the absolute configuration of this sample is the one shown in Figure 5.41a (i.e., type A). Therefore, the diol is (S)-propane-1,2-diol (Figure 5.42f).

The evolution of the ¹H-NMR spectra with temperature is easily explained by consider a Dreiding model of the most stable conformer [63] (i.e., *sp-gt-I*, Figure 5.43), taking into consideration the shielding/deshielding effects experienced by the methylene (1′) protons.

As temperature decreases, the population of the most stable conformer (*sp-gt-I*) increases, as does its contribution to the average NMR spectrum. In this conformation, the high-field methylene proton (i.e., pro-R) is shielded by the carbonyl of the MPA that esterifies the primary alcohol, while the low-field methylene proton (i.e., pro-S) is deshielded. So, at the lower temperature, the proton that resonates at the higher field (pro-R) is slightly more shielded (by 0.03 ppm, which is less than 0.1 ppm), whereas the proton that resonates at the lower field (pro-S) is slightly more deshielded (0.11 ppm).

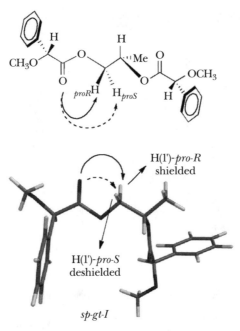

*Figure 5.43.* Main shielding and deshielding effects caused by the carbonyl group of the MPA at the primary alcohol on the methylene (1′) hydrogens in the bis-(R)-MPA ester of (S)-propane-1,2-diol.

### 5.3.9. Example 40: Assignment of the Absolute Configuration of (R)-Propane-1,2-Diol

Figure 5.44 illustrates the application of this simplified method to the assignment of the absolute configuration of (R)-propane-1,2-diol, the diol enantiomeric to the one shown in the previous example.

The $\Delta\delta^{T1T2}$ values for the methylene (1′) hydrogens are +0.25 ppm and −0.05 ppm for the high-field and low-field hydrogens, respectively. A $\Delta\delta^{T1T2}$ larger than 0.10 ppm for the high-field hydrogen indicates that the absolute configuration of the diol is the one shown in Figures 5.41b and 5.44f (i.e., type B).

Inspection of the shielding/deshielding effects on a model of the most stable conformation [63] (i.e., *sp-gt-II*) explains the experimental results (Figure 5.45).

As the NMR probe temperature decreases, the population of the *sp-gt-II* form increases, as does its contribution to the NMR spectrum. In this conformation, the hydrogen that resonates at the higher field (i.e., pro-S) is shielded by the phenyls of the two MPA units and by the carbonyl of the MPA unit at the primary alcohol. Therefore, an increase in the population of this conformer leads to a strong upfield shift of this hydrogen ($\Delta\delta^{T1T2}$ = +0.25 ppm > 0.1 ppm). On the other hand, the hydrogen that resonates at the lower field (i.e., pro-R) is shielded by a phenyl and deshielded by the carbonyl, and as the temperature decreases, it undergoes a slight downfield shift ($\Delta\delta^{T1T2}$ = −0.05 ppm; Figure 5.44d).

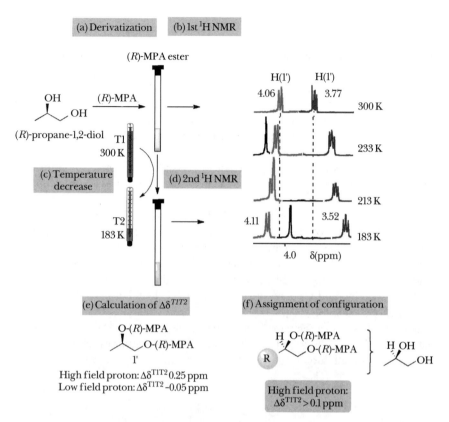

*Figure 5.44.* Main steps in the simplified configurational assignment of (S)-propane-1,2-diol based on the high-field methylene (1′) hydrogen using (R)-MPA.

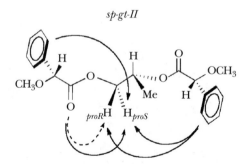

*Figure 5.45.* Main shielding and deshielding effects caused by the MPA carbonyl and phenyl groups on the methylene (1′) hydrogens in the bis-(R)-MPA ester of (R)-propane-1,2-diol.

### 5.3.10. Summary

The assignment of the absolute configuration of *prim/sec*-1,2-diols involves the simultaneous derivatization of the two hydroxyl groups, leading to the corresponding bis-(R)- and bis-(S)-AMAA esters followed by a comparison of their $^1$H-NMR spectra [13, 63–65]. There are two CDAs that can be used for that purpose: MPA [63, 65] and 9-AMA [64]. In both cases, the assignment of the configuration is carried out by considering the $\Delta\delta^{RS}$ signs from both the R substituent and the methylene (1′) proton that has the largest $\Delta\delta^{RS}$ value.

With 9-AMA as the auxiliary reagent, the assignment can be simplified by analyzing the separation between the diastereotopic methylene (1′) protons [64].

In addition to the double-derivatization procedure, *prim/sec*1,2-diols can also be assigned by the low-temperature $^1$H NMR of a single MPA derivative [65]. In this procedure, the evolution with temperature of the signals from the methylene protons is determined by looking at their separation. In practice, it is only necessary to evaluate $\Delta\delta^{T1T2}$ values for the methylene proton that resonates at the higher field and check if its $\Delta\delta^{T1T2}$ value is lower or higher than 0.10 ppm.

To date, no method for assignment based on $^{13}$C NMR has been described for this functional assembly [72].

Figure 5.46 shows a selection of *prim/sec*1,2-diols that have been used to validate the procedures described in this chapter [63–65]. Other examples of applications of this methodology can be found in the literature [215, 216].

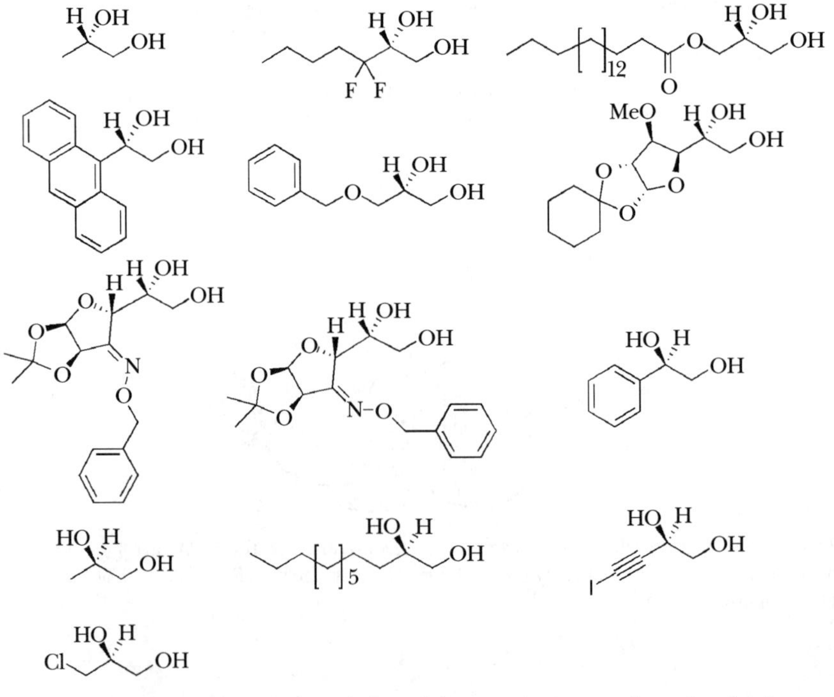

*Figure 5.46.* Selection of *prim/sec*1,2-diols with known absolute configuration that have been used to validate the procedures described in this chapter.

## 5.4. SEC/PRIM-1,2-AMINO ALCOHOLS

The absolute configuration of *sec/prim*-1,2-amino alcohols can be determined by double- [13, 66, 67] and single-derivatization methods [13, 68, 165] using MPA as the CDA (Figure 5.47). The double-derivatization procedure requires the comparison of the $^1$H-NMR spectra of the bis-(R)- and bis-(S)-MPA derivatives. Two sets of signals can be used for the diagnosis: (a) R substituent and methylene (1') hydrogens of the substrate (as in the case of *prim/sec*-1,2-diols), and (b) the OMe and the CαH of the chiral auxiliary (i.e., MPA).

### 5.4.1. Double-Derivatization Methods: MPA and the Use of R and Methylene Hydrogens

Conformational analysis of the bis-(R)- and bis-(S)-MPA derivatives [66, 67] shows that the preference around the Cα–CO bond of the MPA ester moiety is *sp* [37], while that in the MPA amide moiety is *ap* [52]. The C(1')–C(2') and C(1')–O bonds are defined by a number of conformers in equilibrium (*gt-I, gt-III, gg-II,* and *gg-III*; Figure 5.48), of which the rotamer *gt-I* is the most stable. Therefore, the

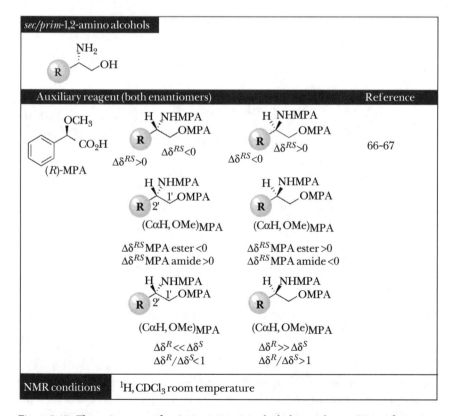

*Figure 5.47.* The assignment of *sec/prim*-1,2-amino alcohols at a glance: CDA, Δδ signs and values (i.e., $\Delta\delta^{RS}$, $\Delta\delta^R$, $\Delta\delta^S$), experimental conditions, and references.

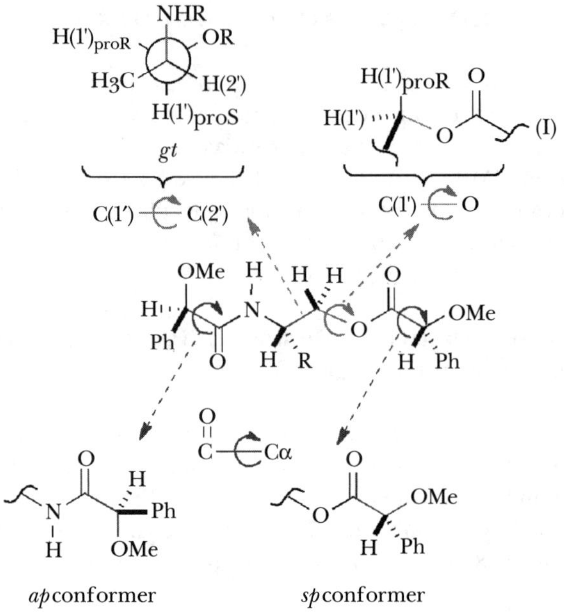

*Figure 5.48.* Main bonds and conformers involved in the conformational equilibria of sec/prim-1,2-amino alcohols.

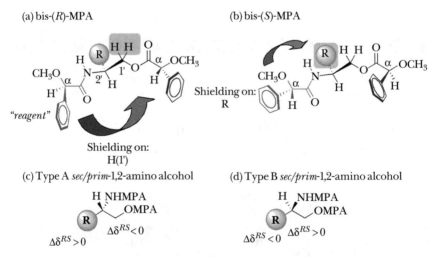

*Figure 5.49.* Main shielding effects and $\Delta\delta^{RS}$ values generated by the MPA phenyl groups on the R and methylene (1′) hydrogens in bis-(R)- and bis-(S)-MPA amido esters of sec/prim-1,2-amino alcohols.

representative structures of the bis-MPA derivatives are well represented by the sp-ap-gt-I conformer, and conversely, their ¹H-NMR spectra and chemical shifts can be predicted considering the shielding and deshielding effects taking place in those conformers [66, 67] (Figure 5.49).

Thus, in a *sec/prim*-1,2-amino alcohol with the absolute configuration shown in Figures 5.49a-c (i.e., type A), the methylene (1′) protons are more shielded in the bis-(*R*)- than in the bis-(*S*)-MPA derivative (Figure 5.49), therefore their $\Delta\delta^{RS}$ values are negative [67] (Figure 5.49c). However, R group protons are more shielded in the bis-(*S*)- than in the bis-(*R*)-MPA derivative, and so their $\Delta\delta^{RS}$ values are positive [67] (Figure 5.49c). The reversed sign distribution is found for enantiomeric amino alcohols [67] (i.e., negative $\Delta\delta^{RS}$ for R protons and positive $\Delta\delta^{RS}$ for methylene ones, $CH_2(1')$; type B; Figure 5.49d).

In this way, the assignment of the absolute configuration of a *sec/prim*-1,2-amino alcohol by double derivatization with MPA requires [67]: (a) comparison of the $^1$H-NMR spectra of the bis-MPA derivatives, (b) examination of the $\Delta\delta^{RS}$ signs for both R and methylene groups, and (c) comparison of that sign distribution with the ones shown in the models in Figures 5.49c and d.

### 5.4.2. Example 41: Assignment of the Absolute Configuration of (*S*)-2-Aminopropan-1-ol Based on *R* and Methylene Hydrogens

A sample of a single enantiomer of 2-aminopropan-1-ol is divided and separately derivatized with (*R*)- and (*S*)-MPA (Figure 5.50a), the resulting bis-MPA derivatives are examined by $^1$H NMR (Figure 5.50b), and the $\Delta\delta^{RS}$ values are calculated for the Me(3′) (+0.14 ppm) and for both methylene hydrogens (−0.12 and −0.09 ppm). The absolute configuration is assigned by matching this distribution of $\Delta\delta^{RS}$ signs with those in Figures 5.49c and d. A positive sign for R [i.e., Me(3′)] and a negative one for the methylene (1′) hydrogens indicate that the absolute configuration is the one shown in Figure 5.49c (type A). Thus, the compound is (*S*)-2-aminopropan-1-ol (Figure 5.50d).

Dreiding models of the bis-MPA derivatives in the most stable conformers (i.e., *sp-ap-gt-I*; Figure 5.51) explain the $^1$H-NMR spectra observed. Methylene (1′) hydrogens are more shielded in the bis-(*R*)- than in the bis-(*S*)-MPA derivative, and their $\Delta\delta^{RS}$ values are negative. The Me(3′) is more shielded in the bis-(*S*)- than in the bis-(*R*)-MPA derivative, and so its $\Delta\delta^{RS}$ is positive.

### 5.4.3. Double-Derivatization Methods: The Use of OMe and C*α*H Signals for Assignment

If the conformations shown in Figures 5.52a and b, corresponding to the *sp-ap-gt-I* conformers of a type A *sec/prim*-1,2-amino alcohol, are analyzed in detail, one can deduce that the shielding cone generated by the phenyl ring of the MPA amide unit in the bis-(*R*)-MPA derivative is pointing towards the C*α*H and OMe of the MPA ester unit [66]. In the bis-(*S*)-MPA derivative, the phenyl ring of the MPA ester shields the C*α*H and OMe of the MPA amide part. Therefore, the $\Delta\delta^{RS}$ of the C*α*H [i.e., $\Delta\delta^{RS}C\alpha H = \delta C\alpha H(R) - \delta C\alpha H(S)$] and that of the OMe [i.e., $\Delta\delta^{RS}OMe = \delta OMe(R) - \delta OMe(S)$] signals from the MPA ester are negative [66], whereas the $\Delta\delta^{RS}$ of the C*α*H and OMe from the MPA amide are positive [66] (type A;

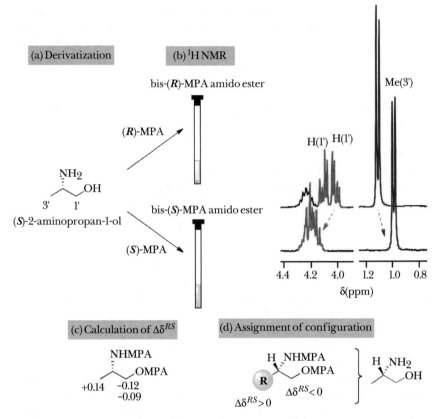

*Figure 5.50.* Main steps in the configurational assignment of (S)-2-aminopropane-1-ol based on R and methylene hydrogens as diagnostic signals using MPA.

Figure 5.52c). The reverse sign distribution is found for enantiomeric amino alcohols [66] (type B; Figure 5.52d).

This conformation (*sp-ap-gt-I*) compels one of the MPA units to act as a reagent, pointing its shielding cone to the other MPA unit that therefore acts as a substrate. The actual shieldings and chemical shifts created by this mutual reagent-substrate relationship depend on the stereochemistry of the bis-MPA derivative, and therefore, the cross-interaction between MPA units allows the use of the OMe and CαH singlets as diagnostic signals for the absolute configuration assignment [66].

In practice, it is not necessary to distinguish the OMe/CαH signals due to the MPA ester unit from those from the MPA amide unit, nor is it necessary to calculate their $\Delta\delta^{RS}$ values, because a direct inspection of the separation of the signals [66], defined as $\Delta\delta^{R}$ and $\Delta\delta^{S}$, is sufficient to establish the correlation between the chemical shifts and the stereochemistry: $\Delta\delta^{R}$CαH = δCαH(low field) − δCαH(high field); $\Delta\delta^{S}$CαH = δCαH(low field) − δCαH(high field); $\Delta\delta^{R}$OMe = δOMe(low field) − δOMe(high field); and $\Delta\delta^{S}$OMe = δOMe(low field) − δOMe(high field). Experimental results [66] have demonstrated that in all the amino alcohols studied,

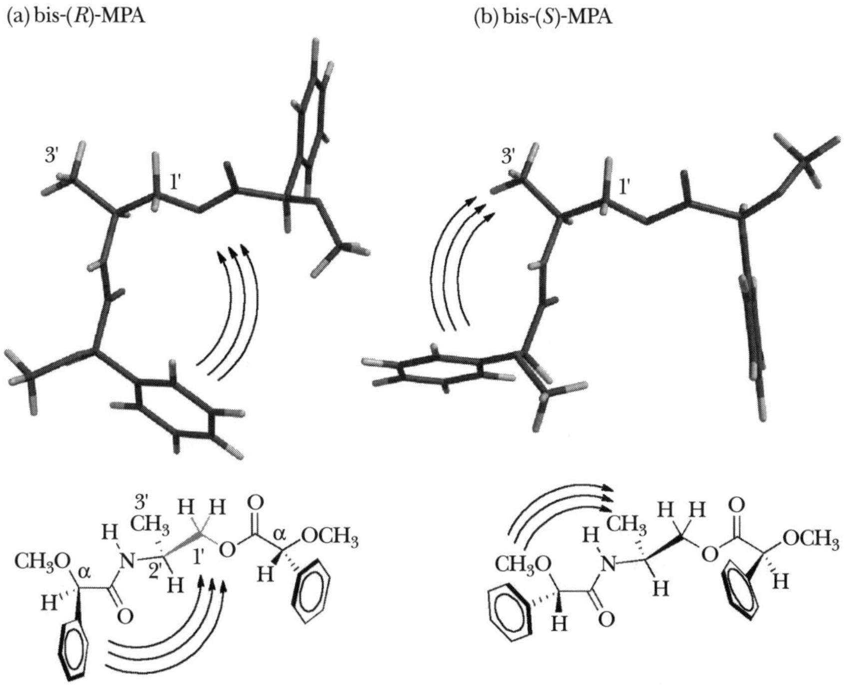

(a) bis-(R)-MPA     (b) bis-(S)-MPA

*Figure 5.51.* Main shielding effects caused by MPA phenyl groups on R and methylene (1′) hydrogens in bis-(R)- and bis-(S)-MPA amido esters of (S)-2-aminopropan-1-ol.

the signals corresponding to the CαH and OMe of the MPA ester are always more deshielded than those of the MPA amide, which makes their assignment in the NMR spectra easier.

Thus, in a *sec/prim*-amino alcohol with the configuration shown in Figures 5.52a and b (type A), the separation between the CαH and OMe signals of the MPA units are shorter in the bis-(R)- than in the bis-(S)-MPA derivative [66] (i.e., $\Delta\delta^R \ll \Delta\delta^S$; Figure 5.52e), while for a *sec/prim*-amino alcohol with the opposite configuration [66] (type B), the opposite holds: $\Delta\delta^R \gg \Delta\delta^S$ (Figure 5.52f).

## 5.4.4. Example 42: Assignment of the Absolute Configuration of (S)-2-Aminopropan-1-ol Using $\Delta\delta^{RS}$ of OMe and CαH Signals

The $^1$H-NMR spectra of the bis-MPA derivatives of an enantiomer of 2-aminopropan-1-ol are shown in Figure 5.53b. Taking into consideration that the signals corresponding to the CαH and OMe of the MPA ester are always more deshielded than those of the MPA amide, the $\Delta\delta^{RS}$CαH and $\Delta\delta^{RS}$OMe values are calculated (Figure 5.53c). Both the CαH and OMe of the MPA amide have positive $\Delta\delta^{RS}$ values (+0.09 and +0.04 ppm, respectively), while those of the MPA ester are negative (−0.12 and −0.06 ppm, respectively). This

160 ■ The Assignment of the Absolute Configuration by NMR

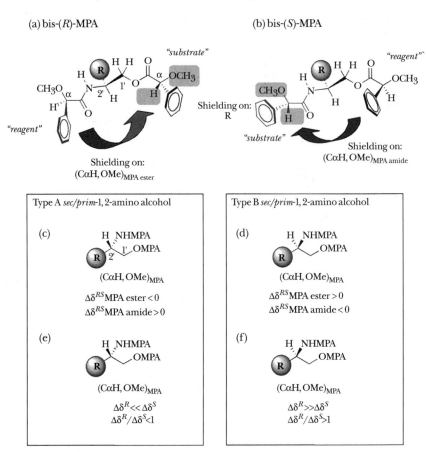

*Figure 5.52.* Main shielding effects, $\Delta\delta^{RS}$, $\Delta\delta^{R}$ and $\Delta\delta^{S}$ values generated by the MPA phenyl groups on the R and methylene (1') hydrogens in bis-(R)- and bis-(S)-MPA amido esters of *sec/prim*-1,2-amino alcohols.

distribution of signs corresponds to the stereochemistry of Figure 5.52c (type A), leading to the absolute configuration shown in Figure 5.53d [i.e., (S)-2-aminopropan-1-ol].

As in many other examples, the spectral differences between the two bis-MPA derivatives are easily explained with Dreiding models representing the *sp-ap-gt-I* conformations and the expected shielding/deshielding effects (Figure 5.54). Thus, in the bis-(R)-MPA derivative, the phenyl ring of the MPA amide unit shields the CαH and OMe of the MPA ester (Figure 5.54a), while in the bis-(S)-MPA derivative, the MPA ester shields the CαH and OMe of the MPA amide (Figure 5.54b). Overall, the $\Delta\delta^{RS}$ values of the CαH/OMe due to the MPA ester must be negative, whereas the $\Delta\delta^{RS}$ of the CαH/OMe at the MPA amide must be positive.

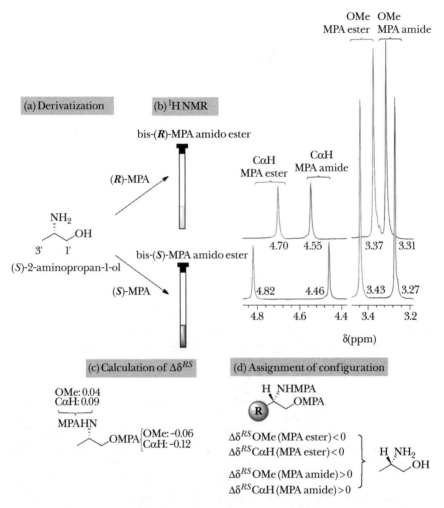

*Figure 5.53.* Main steps in the configurational assignment of (S)-2-aminopropan-1-ol based on OMe and CαH diagnostic signals of the bis-(R)- and bis-(S)-MPA amido esters.

### 5.4.5. Example 43: Assignment of the Absolute Configuration of (S)-2-Aminopropan-1-ol Using the Separation of OMe and CαH Signals

As has been pointed out [66], the assignment of absolute configuration based on OMe and CαH signals does not require precise identification of the OMe and CαH signals from the MPA ester and amide units, nor does it require calculating their $\Delta\delta^{RS}$ values [66]. Visual inspection of the separation of those signals in the bis-(R)- and the bis-(S)-MPA amido esters—defined as $\Delta\delta^R$ and $\Delta\delta^S$, respectively—is sufficient.

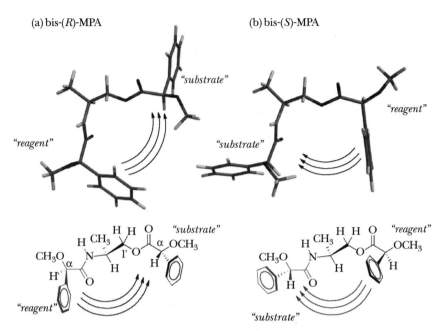

Figure 5.54. Main shielding effects caused by MPA phenyl groups on OMe and CαH groups in bis-(R)- and bis-(S)-MPA amido esters of (S)-2-aminopropan-1-ol.

In this way, the assignment of a sample of an enantiomer of 2-aminopropan-1-ol requires: (a) preparation of the two MPA derivatives, (b) comparison of their $^1$H-NMR spectra, considering only the signals from the CαH and OMe at the two MPA moieties (Figures 5.55a and b), and (c) calculation of the separation between those signals: $\Delta\delta^R$CαH, $\Delta\delta^S$CαH, $\Delta\delta^R$OMe, and $\Delta\delta^S$OMe (absolute values; see Section 5.4.3; Figure 5.55c). The data obtained from the CαH groups are +0.15 and +0.36 ppm for $\Delta\delta^R$CαH and $\Delta\delta^S$CαH, respectively, and those from the OMe groups are +0.06 and +0.16 ppm for $\Delta\delta^R$OMe and $\Delta\delta^S$OMe, respectively. As the separation of CαH and OMe signals are smaller in the bis-(R)- than in the bis-(S)-MPA derivative ($\Delta\delta^R \ll \Delta\delta^S$), the configuration of the amino alcohol corresponds to that shown in Figures 5.52e and 5.55d (type A).

The shift differences in the spectra can be explained by building Dreiding models of the bis-MPA derivatives placed in the *sp-ap-gt-I* conformation [66] (Figure 5.54). In the bis-(R)-MPA derivative, the CαH and OMe of the MPA ester unit are shielded, while those of the MPA amide are not affected (Figures 5.54a and 5.55b). In the bis-(S)-MPA derivative, the CαH and OMe of the MPA amide are shielded, while those of the MPA ester are not affected (Figures 5.54b and 5.55b). Overall, the separation ($\Delta\delta$) between the signals of the two CαH is smaller in the bis-(R)-MPA derivative (i.e., $\Delta\delta^R$CαH) than it is in the bis-(S)-MPA derivative (i.e., $\Delta\delta^S$CαH). Analogous results are obtained for the OMe groups [66] (i.e., $\Delta\delta^R$OMe is smaller than $\Delta\delta^S$OMe).

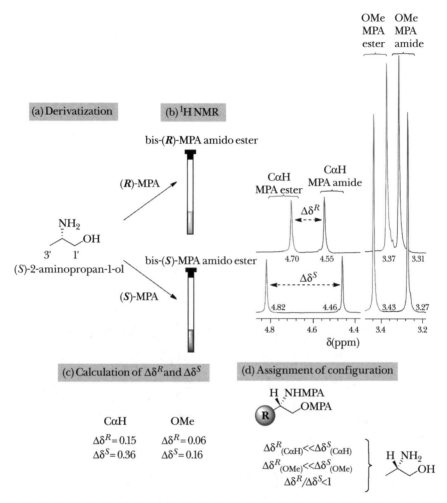

*Figure 5.55.* Main steps in the configurational assignment of (S)-2-aminopropan-1-ol based on the separation of OMe and CαH diagnostic signals of the bis-(R)- and bis-(S)-MPA amido esters.

### 5.4.6. Example 44: Assignment of the Absolute Configuration of (R)-2-Amino-3-Methylbutan-1-ol Using the Separation of OMe and CαH Signals

A sample of 2-amino-3-methylbutan-1-ol is divided and separately derivatized with (R)- and (S)-MPA (Figure 5.56a). The ¹H-NMR spectra of the resulting bis-MPA derivatives are registered, and the zone of the CαH and OMe signals is analyzed [66] (Figure 5.56b). It is not necessary to identify the exact origin of the OMe and CαH signals (i.e., if they originated in the MPA ester or the MPA amide), because just the separation between them [i.e., between the two CαH ($\Delta\delta^R$CαH, $\Delta\delta^S$CαH) and between the two OMe ($\Delta\delta^R$OMe, $\Delta\delta^S$OMe) signals in

164 ■ The Assignment of the Absolute Configuration by NMR

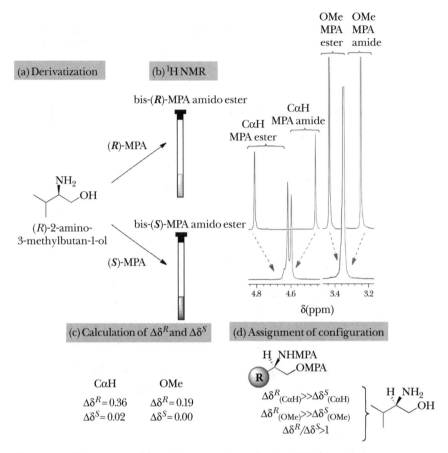

*Figure 5.56.* Main steps in the configurational assignment of (R)-2-amino-3-methylbutan-1-ol based on the separation of OMe and CαH diagnostic signals of the bis-(R)- and bis-(S)-MPA amido esters.

each bis-MPA amido ester] are sufficient for assignment. The values obtained are +0.36 ppm and +0.02 ppm ($\Delta\delta^R$ and $\Delta\delta^S$, respectively) for CαH, and +0.19 and 0.00 ppm ($\Delta\delta^R$ and $\Delta\delta^S$, respectively) for OMe. As the separations of CαH and OMe are larger in the bis-(R)- than in the bis-(S)-MPA derivative ($\Delta\delta^R \gg \Delta\delta^S$), the configuration of the amino alcohol corresponds to that shown in Figures 5.52f and 5.56d [type B; (R)-2-amino-3-methylbutan-1-ol].

These observations of the $^1$H-NMR spectra can be explained by building Dreiding models of the bis-MPA derivatives placed in the *sp-ap-gt-I* conformation (Figure 5.57). In the bis-(R)-MPA derivative, the CαH and OMe groups of the MPA amide are shielded, while those of the MPA ester are not affected (Figures 5.57a and 5.56b). In the bis-(S)-MPA derivative, the CαH and OMe groups of the MPA ester are shielded, while those of the MPA amide are not affected (Figures 5.57b and 5.56b). Overall, the separation ($\Delta\delta$) between the signals of the two CαH is larger in the bis-(R)-MPA derivative ($\Delta\delta^R$CαH) than in the bis-(S)-MPA derivative

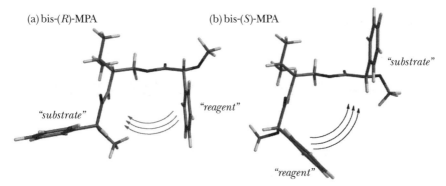

*Figure 5.57.* Main shielding effects caused by MPA phenyl groups on OMe and CαH groups in bis-(R)- and bis-(S)-MPA amido esters of (R)-2-amino-3-methylbutan-1-ol.

($\Delta\delta^S$CαH). Similar results are observed with the OMe signals ($\Delta\delta^R$OMe is larger than $\Delta\delta^S$OMe).

### 5.4.7. Single-Derivatization Method: MPA

The absolute configuration of *sec/prim*-1,2-amino alcohols can also be assigned by low-temperature NMR of only one bis-MPA derivative by using the CαH protons of the two MPA units as the diagnostic signals [13, 68, 165] (Figure 5.58).

The bis-MPA derivatives of these amino alcohols consist mainly of a conformational equilibrium between the two main conformers at the MPA moieties (Figure 5.59a): the most stable one is represented by the *sp* conformation in the MPA ester unit and the *ap* conformation in the MPA amide unit (i.e., *sp-ap-gt-I*; for a full description of all the conformations, see Section 5.4.1). The second most stable conformation is represented by the *ap* conformation in the MPA ester unit and the *sp* in the MPA amide unit [66, 68].

A decrease in temperature leads to an increase in the population of molecules in the most stable conformation and a corresponding increase in its contribution to the average NMR spectrum [41, 68]. Thus, the evolution with temperature of the NMR spectra of the bis-MPA derivatives can be easily predicted on the basis of the two most relevant conformations, which are shown in Figure 5.59a.

In the most stable conformer of the bis-(R)-MPA derivative of a type A *sec/prim*-1,2-amino alcohol (*sp-ap-gt-I*; Figure 5.59a), the phenyl ring of the *ap* MPA amide group shields the protons contained in the *sp* MPA ester [66, 68]. These reagent-substrate roles are reversed in the least stable conformation: the *ap* MPA ester unit shields the signals of the *sp* MPA amide unit [66, 68].

A decrease in temperature produces an increase in the number of molecules in which the MPA ester protons are shielded and a decrease in the number of molecules in which the MPA amide signals are shielded. Overall, lowering the temperature compels the signal due to the CαH of the MPA ester to shift upfield [68] (i.e., positive $\Delta\delta^{T1T2}$) and that from the MPA amide to shift downfield [68] (i.e., negative $\Delta\delta^{T1T2}$; Figure 5.59b).

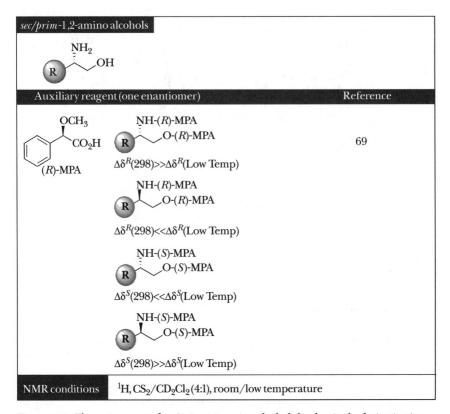

*Figure 5.58.* The assignment of *sec/prim*-1,2-amino alcohols by the single-derivatization method at a glance: CDA, diagnostic shifts ($\Delta\delta^R$, $\Delta\delta^S$), experimental conditions, and references.

As in the case of the double-derivatization methods shown previously (Section 5.4.3), it is not necessary to distinguish the CαH signal due to the MPA ester from that produced by the MPA amide; it is sufficient to evaluate the separation between the two CαH ($\Delta\delta$) at different temperatures [68]. For the stereochemistry shown in Figure 5.59 (type A), the separation of the CαH signals is larger at room temperature than it is at a lower temperature [68] [i.e., $\Delta\delta^R(298) \gg \Delta\delta^R$(low temp); Figures 5.59b and c].

If the bis-(*S*)-MPA derivative of the same amino alcohol is analyzed, the evolution with temperature is the opposite [68] [i.e., $\Delta\delta^S(298) \ll \Delta\delta^S$(low temp); Figure 5.59d].

Analogous reasoning applied to the bis-(*R*)-MPA derivative of the enantiomeric alcohol (type B) shows that the CαH of the MPA amide is shielded in the most abundant conformer, whereas in the least stable one, the MPA ester unit is the one that is shielded (Figure 5.60a). As the temperature decreases, the CαH signal of the MPA ester is more deshielded (i.e., negative $\Delta\delta^{T1T2}$; Figure 5.60b), while the CαH of the MPA amide is more shielded (i.e., positive $\Delta\delta^{T1T2}$; Figure 5.60b). As a result, the separation between the CαH signals is smaller at

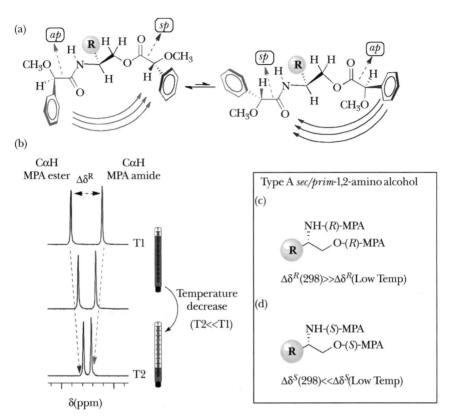

*Figure 5.59.* Simplified procedure for assignment of type A *sec/prim*-1,2-amino alcohols based on $\Delta\delta^R$ and $\Delta\delta^S$ parameters from CαH at different temperatures in bis-MPA esters.

room temperature than it is at a lower temperature [68] [i.e., $\Delta\delta^R(298) \ll \Delta\delta^R$(low temp); Figures 5.60b and c].

Similar reasoning applied to the bis-(S)-MPA derivative of the same amino-alcohol leads to the opposite result [68] [i.e., $\Delta\delta^S(298) \gg \Delta\delta^S$(low temp); Figure 162d].

### 5.4.8. Example 45: Assignment of the Absolute Configuration of (S)-2-Aminopropan-1-ol by Low-Temperature NMR of a Single Derivative

A sample of 2-aminopropan-1-ol is converted into the bis-(R)-MPA derivative (Figure 5.61a), and its ¹H-NMR spectra is registered at 298 and 183 K (Figure 5.61b), focusing on the spectral zone of the CαH singlets. The separation between the two CαH signals is calculated at room temperature [$\Delta\delta^R(298) = +0.17$ ppm] and at a lower temperature [$\Delta\delta^R(183) = +0.12$ ppm; Figure 5.61e)].

As $\Delta\delta^R(298) \gg \Delta\delta^R$(low temp), the absolute configuration of the amino alcohol is the one shown in Figures 5.59c and 5.61f (type A; (S)-2-aminopropan-1-ol).

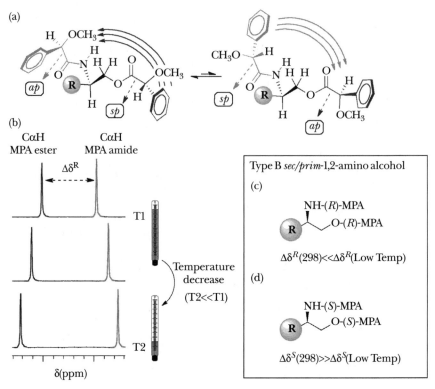

*Figure 5.60.* Simplified procedure for assignment of type B *sec/prim*-1,2-amino alcohols based on $\Delta\delta^R$ and $\Delta\delta^S$ parameters from CαH at different temperatures in bis-MPA esters.

These results can be easily explained by building a Dreiding model of the bis-(R)-MPA derivative of (S)-2-aminopropan-1-ol in the most stable conformer, *sp-ap-gt-I* (Figure 5.62). In this conformation, the CαH of the MPA ester is shielded by the MPA amide unit (the signals from the MPA amide are shielded in the least stable conformer). A lower temperature increases the population of the *sp-ap-gt-I* conformer, and on average, the signals of the MPA ester become more shielded, and those of the MPA amide become deshielded. The result is that the separation between the CαH singlets is larger at room than it is at low temperature.

### 5.4.9. Example 46: Assignment of the Absolute Configuration of (R)-2-Aminopropan-1-ol by Low-Temperature NMR of a Single Derivative

A small sample of an enantiomer of 2-aminopropan-1-ol is derivatized with (R)-MPA (Figure 5.63a), and its $^1$H-NMR spectra is registered at 298 and 183 K (Figure 5.63b). The separation between the CαH singlets at room temperature and a lower temperature is calculated [$\Delta\delta^R(298)$ = +0.37 ppm and $\Delta\delta^R(183)$ = +0.47 ppm].

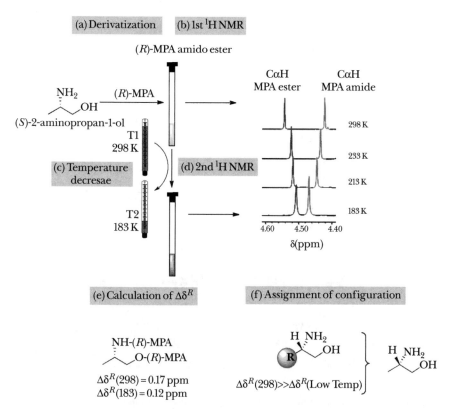

Figure 5.61. Main steps in the simplified configurational assignment of (S)-2-aminopropan-1-ol based on CαH diagnostic signals from the bis-(R)-MPA amido ester at different temperatures.

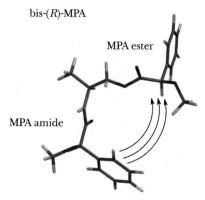

Figure 5.62. Main shielding effects caused by MPA phenyl groups in the bis-(R)-MPA amido ester of (S)-2-aminopropan-1-ol.

# 170 ■ The Assignment of the Absolute Configuration by NMR

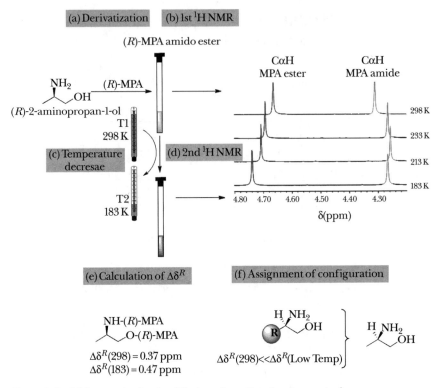

Figure 5.63. Main steps in the simplified configurational assignment of (R)-2-aminopropan-1-ol based on CαH diagnostic signals from the bis-(R)-MPA amido ester at different temperatures.

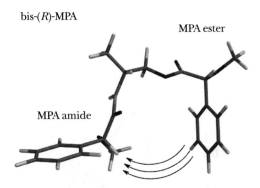

Figure 5.64. Main shielding effects caused by MPA phenyl groups in the bis-(R)-MPA amido ester of (R)-2-aminopropan-1-ol.

As $\Delta\delta^R(298) \ll \Delta\delta^R$(low temp), the absolute configuration of the amino alcohol is the one shown in Figure 5.60c [type B; (R)-2-aminopropan-1-ol; Figure 5.63f].

Explanation of these results can be obtained by building a Dreiding model of the bis-(R)-MPA derivative of (R)-2-aminopropan-1-ol (Figure 5.64). When the most stable conformer is considered (*sp-ap-gt-I*), the CαH of the MPA amide unit is shielded by the MPA ester unit. Lowering the temperature increases the population of that conformer and, on average, this signal becomes more shielded. In the least stable conformer, the CαH of the MPA ester is shielded, and at lower temperatures, it shifts downfield. The overall consequence is that the separation between the CαH singlets is larger at low temperatures than at room temperature [68].

## 5.4.10. Summary

The absolute configuration of *sec/prim*-1,2-amino alcohols can be determined by double and single-derivatization methods [13, 66–68, 165] using MPA as the CDA in both cases. The double-derivatization procedure [66, 67] requires the comparison of the $^1$H-NMR spectra of the bis-(R)- and bis-(S)-MPA derivatives by using the $\Delta\delta^{RS}$ signs from the R and methylene (1′) protons, as in the case of *prim/sec*-1,2-diols. An alternative is to focus not on the signals from the protons of the substrate amino alcohol, but on the OMe and CαH signals of the MPA moieties.

These signals can be used for assignment in two ways: (a) using the $\Delta\delta^{RS}$ signs of the OMe and CαH of both MPA units, and (b) by comparison of the separation between the CαH and OMe signals of the two MPA units in the bis-(R)- and bis-(S)-MPA derivatives ($\Delta\delta^R$ and $\Delta\delta^S$, respectively).

In addition, low-temperature $^1$H NMR of a single MPA derivative can also be used for the assignment of these amino alcohols by focusing on the evolution with temperature of the signals due to the CαH protons of the two MPA units [68] (MPA amide and MPA ester) in only one bis-MPA derivative, either the bis-(R) or the bis-(S). In practice, we just need to look at the separation between those two singlets at room temperature and at a lower temperature and observe if $\Delta\delta^R(298)$ is smaller or larger than $\Delta\delta^R$(low temp), and similarly with $\Delta\delta^S(298)$ and $\Delta\delta^S$(low temp).

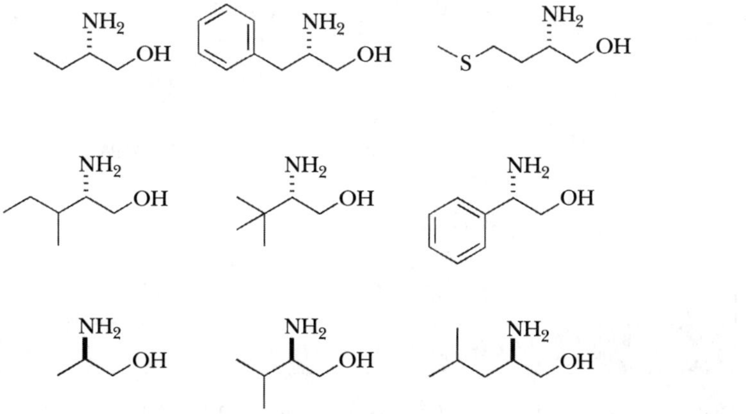

*Figure 5.65.* Selection of *sec/prim*-1,2-amino alcohols with known absolute configuration that have been used to validate the procedures described in this chapter.

To date, no procedure for assignment of this class of compounds has been demonstrated based on $^{13}$C NMR [72].

Figure 5.65 shows a selection of *sec/prim*-1,2-amino alcohols that have been used to validate the procedures described above [66–68]. Other examples of applications of this methodology can be found in the literature [217–220].

## 5.5. PRIM/SEC-1,2-AMINO ALCOHOLS

As in the previous case, the absolute configuration of *prim/sec*-1,2-amino alcohols can be determined by double- [13, 66, 67] and by single-derivatization [13, 68, 165] using MPA as the auxiliary reagent (Figure 5.66). The double-derivatization method requires the preparation of the bis-(R)- and bis-(S)-MPA derivatives and comparison of their $^1$H-NMR spectra. Once again, two sets of NMR signals can be used for diagnosis: (a) the R and methylene (1′) protons, and (b) the OMe and CαH of the MPA units.

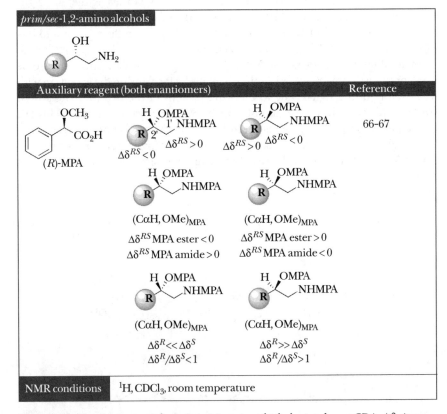

Figure 5.66. The assignment of *prim/sec*-1,2-amino alcohols at a glance: CDA, Δδ signs and values (i.e., $\Delta\delta^{RS}$, $\Delta\delta^R$, $\Delta\delta^S$), experimental conditions, and references.

## 5.5.1. Double-Derivatization Methods: MPA and the Use of R and Methylene Hydrogens

Conformational analysis (Figure 5.67) of the two derivatives [66, 67] [i.e., bis-(R)- and bis-(S)-MPA amido esters] shows that the Cα–CO bond of the MPA ester is represented by an *sp/ap* equilibrium in which the *sp* is the major conformer [37]. In the MPA amide, the *ap* conformer [52] is the most stable one. As for the other bonds involved [i.e., C(1′)–C(2′) and C(1′)–N], the analysis indicates the presence of the following conformers in equilibrium: in the bis-(R)-MPA derivative, two main forms predominate: *gt-II* and *gt-I*, with *gt-II* being the most stable, while the bis-(S)-MPA derivative has three main conformers: *gg-I*, *gt-II*, and *gt-I*, with *gg-I* being the most stable (Figure 5.67).

Thus, the NMR characteristics of the bis-MPA derivatives can be easily predicted by analysis of the shielding and deshielding effects transmitted by the MPA units placed in the most representative conformers [66, 67] (Figure 5.68).

In a *prim/sec*-1,2-amino alcohol with the absolute configuration shown in Figures 5.68a and b (type A), the R group is more shielded in the bis-(R)- than in the bis-(S)-MPA derivative, while the methylene (1′) protons [CH$_2$(1′)] are more shielded in the bis-(S)- than in the bis-(R)-MPA derivatives [67]. The overall result is a negative $\Delta\delta^{RS}$ value for the R protons and a positive one for the methylene (1′) protons (type A; Figure 5.68c). The enantiomeric amino alcohols produce opposite $\Delta\delta^{RS}$ signs [67]: positive $\Delta\delta^{RS}$ for the R protons and negative for the methylene (1′)

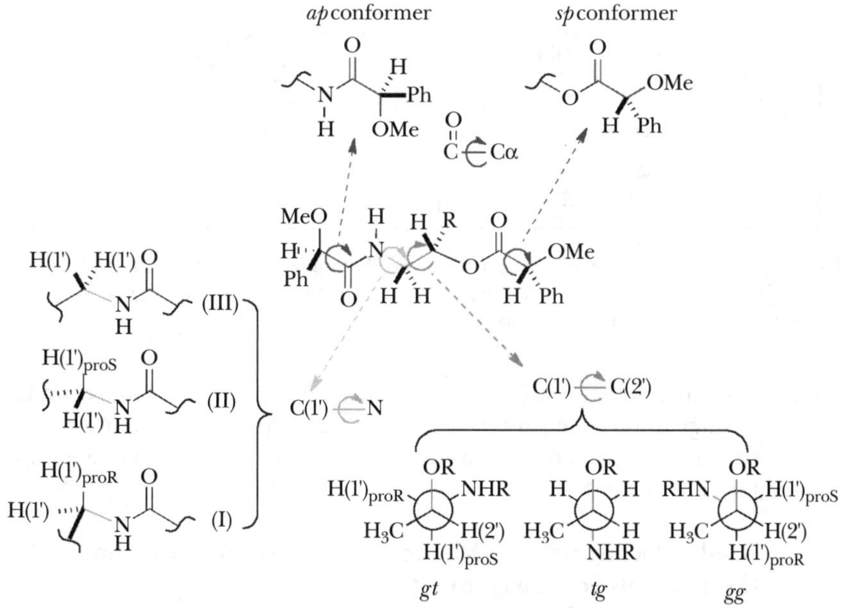

*Figure 5.67.* Main bonds and conformers involved in the conformational equilibria of *prim/sec*-1,2-amino alcohols.

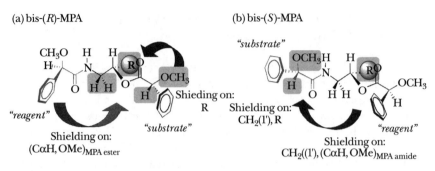

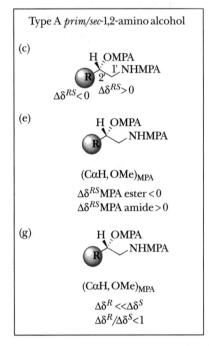
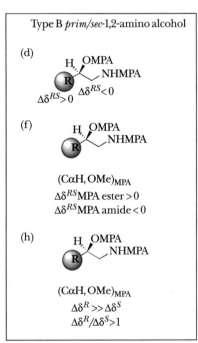

*Figure 5.68.* Main shielding effects, $\Delta\delta^{RS}$, $\Delta\delta^{R}$ and $\Delta\delta^{S}$ values generated by the MPA phenyl groups on the R and methylene (1′) hydrogens in bis-(R)- and bis-(S)-MPA amido esters of *prim/sec*-1,2-amino alcohols.

protons (type B, Figure 5.68d). This correlation between the $\Delta\delta^{RS}$ signs and the absolute configuration has been experimentally proven with a number of structurally varied amino alcohols of known absolute configuration, and therefore it can be used for assignment (Section 5.5.10).

### 5.5.2. Double-Derivatization Methods: The Use of OMe and CαH Signals for Assignment

Furthermore, Figure 5.68 shows that, similarly to what had been observed in *sec/prim*-1,2-amino alcohols [66] (Section, 5.4.3), there is also a cross shielding between the MPA units: in the bis-(R)-MPA derivative, the phenyl ring of the MPA amide unit points its shielding cone towards the CaH and OMe of the MPA ester unit, while in the bis-(S)-MPA derivative, the MPA ester unit shields the CaH and OMe of the MPA amide unit. As a consequence, for the type A stereochemistry shown in Figures 5.68a and b, the $\Delta\delta^{RS}$ signs of the CaH and OMe of the MPA ester unit are negative [66], whereas the $\Delta\delta^{RS}$ signs of the CaH and OMe of the MPA amide unit are positive [66] (Figure 5.68e). The opposite signs are found for the enantiomeric amino alcohols [66] (type B; Figure 5.68f). As before, this correlation between the $\Delta\delta^{RS}$ sign and the absolute configuration has been experimentally proven with a collection of known amino alcohols, and therefore it can be used for assignment [66].

Again, as in the case of *sec/prim*-1,2-amino alcohols [66] (Section 5.4.3), it is not necessary to identify and distinguish between the OMe and CaH signals due to the MPA ester and amide units, nor is it necessary to calculate their $\Delta\delta^{RS}$ values, because the direct inspection of the separation of the signals [66] in the spectra of the bis-(R)- and the bis-(S)-MPA amido esters [i.e., $\Delta\delta^{R}$ and $\Delta\delta^{S}$, respectively; $\Delta\delta^{R}$CaH = $\delta$CaH(low field) – $\delta$CaH(high field); $\Delta\delta^{S}$CaH = $\delta$CaH(low field) – $\delta$CaH(high field); $\Delta\delta^{R}$OMe = $\delta$OMe(low field) – $\delta$OMe(high field); and $\Delta\delta^{S}$OMe = $\delta$OMe(low field) – $\delta$OMe(high field)] is sufficient to establish the correlation between the chemical shifts and the absolute stereochemistry. For a *prim/sec*-1,2-amino alcohol with the configuration shown in 5.68a (i.e., type A), the separation of CaH and OMe signals of the MPA units are smaller in the bis-(R)- than in the bis-(S)-MPA derivative [66]: $\Delta\delta^{R} \ll \Delta\delta^{S}$ (Figure 5.68g), while a *sec/prim*-amino alcohol with the opposite configuration (i.e., type B) shows the opposite NMR behavior [66] ($\Delta\delta^{R} \gg \Delta\delta^{S}$; Figure 5.68h).

### 5.5.3. Example 47: Assignment of the Absolute Configuration of (S)-1-Aminopropan-2-ol Based on R and Methylene Hydrogens

A sample of 1-aminopropan-2-ol is divided and separately derivatized with (R)- and (S)-MPA (Figure 5.69a), the ¹H-NMR spectra of the resulting bis-MPA derivatives are registered (Figure 5.69b), and the $\Delta\delta^{RS}$ values are calculated for Me(3′) (−0.18 ppm) and for both methylene (1′) protons (+0.10 ppm; Figure 5.59c). The absolute configuration is assigned by comparison of this distribution of $\Delta\delta^{RS}$ signs with those shown in Figures 5.68c and d. In this example, a negative sign for the R [i.e., Me(3′)] and a positive one for the methylene (1′) protons indicates the absolute configuration shown in Figure 5.68c (type A), that is, (S)-1-aminopropan-2-ol (Figure 5.69d).

These ¹H-NMR differences between the two bis-MPA derivatives can be structurally visualized and explained by considering Dreiding models of the two bis-MPA derivatives fixed in the most representative conformations (*ap-sp-gt-II* and *ap-sp-gg-I*, Figure 5.70). In the bis-(R)-MPA derivative, Me(3′) is shielded by the MPA ester (Figure 5.70a), while the methylene (1′) protons are not affected. In the bis-(S)-MPA derivative, the methylene (1′) protons are shielded by the MPA ester, but Me(3′) is not affected (Figure 5.70b). Overall, $\Delta\delta^{RS}$ is negative for Me(3′) and positive for H(1′).

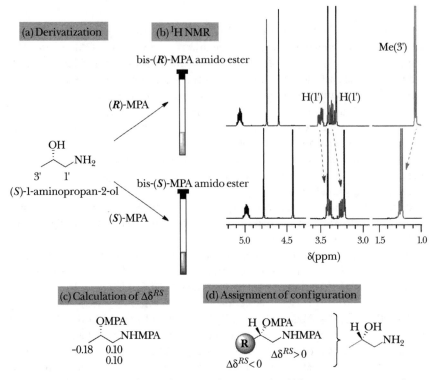

*Figure 5.69.* Main steps in the configurational assignment of (S)-1-aminopropan-2-ol based on R and methylene hydrogens as diagnostic signals using MPA.

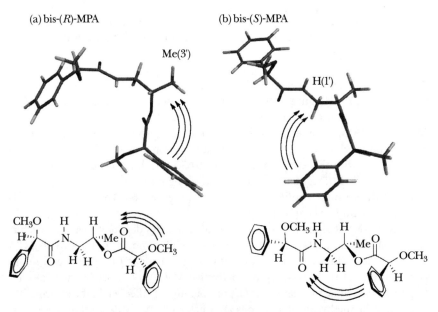

*Figure 5.70.* Main shielding effects caused by MPA phenyl groups on R and methylene (1′) hydrogens in bis-(R)- and bis-(S)-MPA amido esters of (S)-1-aminopropan-2-ol.

## 5.5.4. Example 48: Assignment of the Absolute Configuration of (S)-1-Aminopropan-2-ol Using $\Delta\delta^{RS}$ of OMe and CαH Signals

The bis-MPA derivatives of the previous example are now analyzed focusing on the NMR signals due to the CαH and OMe protons [66]. Their $\Delta\delta^{RS}$ values (Figure 5.71c) are positive for the CαH and OMe of the MPA amide (+0.16 and +0.08 ppm respectively) and negative (−0.04 and −0.02 ppm) for the MPA ester unit.

Comparison of this distribution of signs with Figure 5.68e (type A) indicates that the *prim/sec*-1,2-amino alcohol has the configuration shown in Figure 5.71d, namely, (S)-1-aminopropan-2-ol.

Consideration of the most representative conformation (Figure 5.72) and the orientation of the phenyl rings allow the understanding of the NMR spectra and the shieldings that are observed experimentally. In the bis-(R)-MPA derivative, the MPA amide unit shields the CαH and OMe of the MPA ester (Figure 5.72a), while

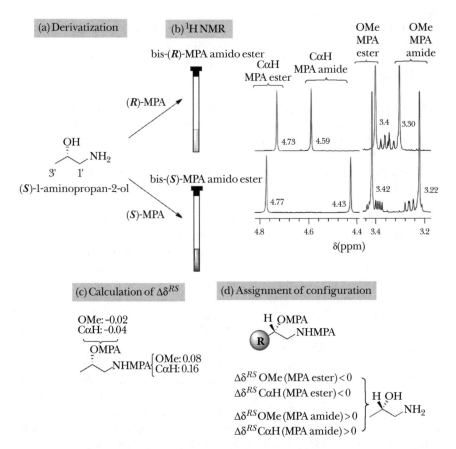

Figure 5.71. Main steps in the configurational assignment of (S)-1-aminopropan-2-ol based on OMe and CαH diagnostic signals of the bis-(R)- and bis-(S)-MPA amido esters.

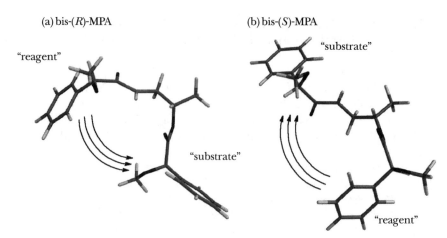

*Figure 5.72.* Main shielding effects caused by MPA phenyl groups on OMe and CαH groups in bis-(R)- and bis-(S)-MPA amido esters of (S)-1-aminopropan-2-ol.

in the bis-(S)-MPA derivative, the MPA ester unit shields the CαH and OMe of the MPA amide (Figure 5.72b). As a consequence, for that stereochemistry, $\Delta\delta^{RS}$ values of CαH/OMe of the MPA ester are negative, whereas those of CαH/OMe of the MPA amide are positive.

### 5.5.5. Example 49: Assignment of the Absolute Configuration of (S)-1-Aminopropan-2-ol Using the Separation of the OMe and CαH signals

As shown before (Section 5.5.2), to assign the absolute configuration, it is not necessary to identify the OMe and CαH signals from the MPA ester and MPA amide, nor is it necessary to calculate their $\Delta\delta^{RS}$ values [66]. Calculation of the separation of their signals in the bis-(R)- and the bis-(S)-MPA derivatives (i.e., $\Delta\delta^R$ and $\Delta\delta^S$, respectively) is sufficient for assignment (see Section 5.5.2).

The separation between OMe and CαH signals in the bis-(R)- and in the bis-(S)-MPA derivatives ($\Delta\delta^R$ and $\Delta\delta^S$; absolute values; Figure 5.73c) are calculated to be +0.14 and +0.34 ppm for $\Delta\delta^R$CαH and $\Delta\delta^S$CαH, respectively, for the CαH, and +0.10 and +0.20 ppm for $\Delta\delta^R$OMe and $\Delta\delta^S$OMe, respectively, for the OMe. As the separations between CαH and OMe signals are smaller in the bis-(R)- than in the bis-(S)-MPA derivative ($\Delta\delta^R \ll \Delta\delta^S$), the configuration of the amino alcohol is the one shown in Figures 5.68g and 5.73d (type A).

Figure 5.72 illustrates graphically how the shielding effects produced in one MPA unit are transmitted to the other. In the bis-(R)-MPA derivative, the CαH and OMe of the MPA ester are shielded by the phenyl ring of the MPA amide (Figures 5.72a and 5.73b), while in the bis-(S)-MPA derivative, the CαH and OMe of the MPA amide are shielded by the phenyl ring of the MPA ester (Figures 5.72b and 5.73b). Therefore, the resulting separation ($\Delta\delta$) between the signals of the two CαH is smaller in the bis-(R)-MPA ($\Delta\delta^R$CαH) than in the

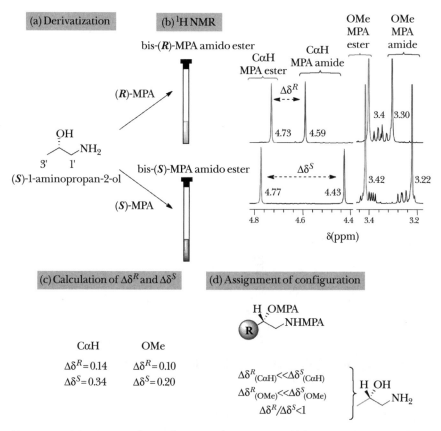

*Figure 5.73.* Main steps in the configurational assignment of (S)-1-aminopropan-2-ol based on the separation of OMe and CαH diagnostic signals of the bis-(R)- and bis-(S)-MPA amido esters.

bis-(S)-MPA derivative ($\Delta\delta^S$CαH). This is also found for the OMe signals and the $\Delta\delta^R$OMe and $\Delta\delta^S$OMe values [66].

### 5.5.6. Example 50: Assignment of the Absolute Configuration of (R)-1-Aminoheptan-2-ol Using the Separation of the OMe and CαH Signals

A sample of 1-aminoheptan-2-ol is divided and separately derivatized with (R)- and (S)-MPA (Figure 5.74a). The $^1$H-NMR spectra of the resulting bis-MPA derivatives are registered, the NMR zone of the CαH and OMe signals is analyzed (Figure 5.74b), and the separation of those signals in the bis-(R)- and the bis-(S)-MPA ($\Delta\delta^R$ and $\Delta\delta^S$, respectively) are evaluated [66]. The values obtained experimentally are $\Delta\delta^R$ = +0.40 ppm and $\Delta\delta^S$ = +0.12 ppm for the CαH (i.e., $\Delta\delta^R$CαH and $\Delta\delta^S$CαH) and $\Delta\delta^R$ = +0.21 and $\Delta\delta^S$ = +0.07 ppm for the OMe (i.e.,

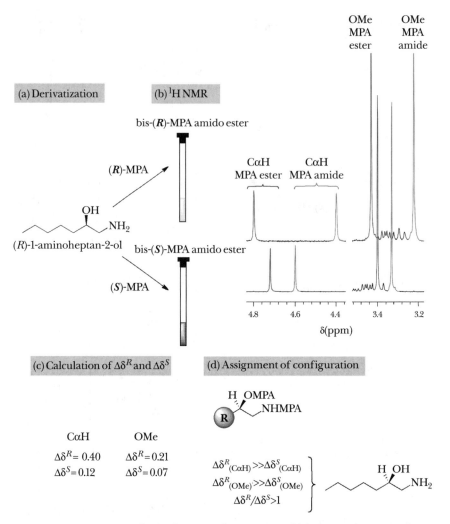

Figure 5.74. Main steps in the configurational assignment of (R)-1-aminoheptan-2-ol based on the separation of OMe and CαH diagnostic signals of the bis-(R)- and bis-(S)-MPA amido esters.

$\Delta\delta^R$OMe and $\Delta\delta^S$OMe). As the separations between the CαH and the OMe signals are smaller in the bis-(S)- than in the bis-(R)-MPA derivative ($\Delta\delta^R \gg \Delta\delta^S$), the configuration of the amino alcohol corresponds to that shown in Figures 5.68h and 5.74d [i.e., type B; (R)-1-aminoheptan-2-ol].

These shifts can be explained by building a Dreiding stereomodel of the bis-(R)- and bis-S-MPA derivatives in the most representative conformation (Figure 5.75). Analysis of the orientation of the anisotropic phenyl rings shows that in the bis-(R)-MPA derivative, the CαH and OMe of the MPA amide unit are shielded by the phenyl ring of the MPA ester (Figures 5.75a and 5.74b). On the other hand, in the bis-(S)-MPA derivative, the CαH and OMe of the MPA ester are shielded by

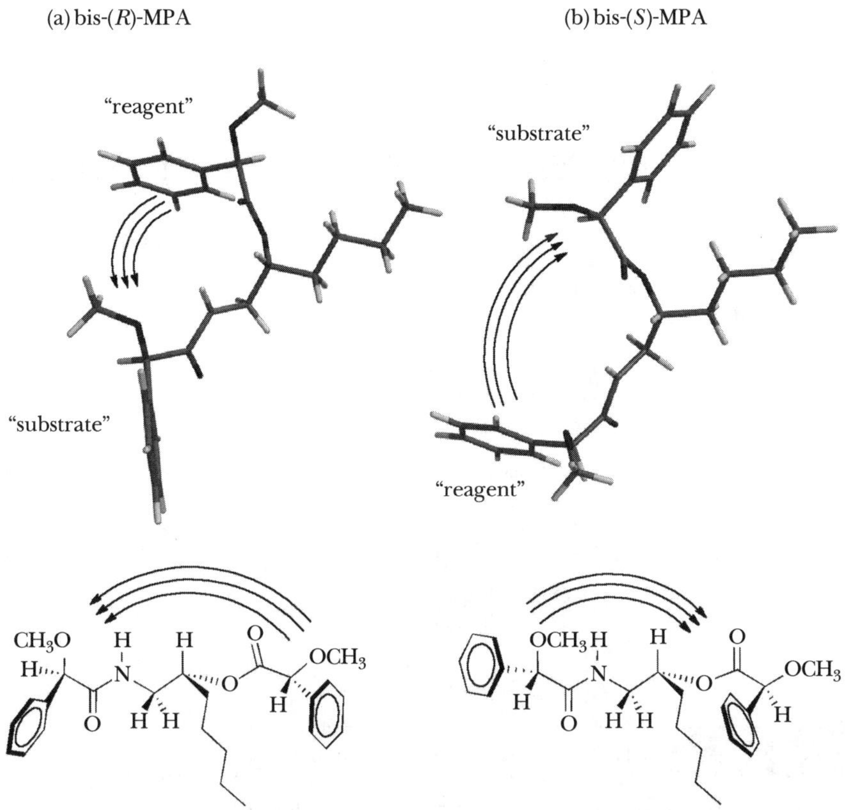

(a) bis-(*R*)-MPA       (b) bis-(*S*)-MPA

*Figure 5.75.* Main shielding effects caused by MPA phenyl groups on OMe and CαH groups in bis-(*R*)- and bis-(*S*)-MPA amido esters of (*R*)-1-aminoheptan-2-ol.

the MPA amide (Figures 5.75b and 5.74b). The experimental consequence of those effects is that the separations (Δδ) between the two CαH signals (i.e., $\Delta\delta^R$CαH and $\Delta\delta^S$CαH) and between the two OMe (i.e., $\Delta\delta^R$OMe and $\Delta\delta^S$OMe) are shorter in the bis-(*S*)-MPA derivative ($\Delta\delta^S$) than in the bis-(*R*)-MPA derivative ($\Delta\delta^R$).

### 5.5.7. Single-Derivatization Method: MPA

As in the case of *sec/prim*-1,2-amino alcohols (Section 5.4.7), it is possible to determine the absolute configuration of a *prim/sec*-1,2-amino alcohol using low-temperature NMR of one derivative only [13, 68, 165] (Figure 5.76). This procedure is carried out by analyzing the evolution with temperature of the signals of the CαH protons of the MPA auxiliaries [13, 68, 165].

The conformational equilibrium of the bis-MPA derivatives of these amino alcohols consists of two main conformers: in the most stable one, the MPA ester unit is in the *sp* conformation [37] and the MPA amide unit is in the *ap* [52], whereas in the second most stable one, the MPA ester is in the *ap* conformation and the MPA amide is in the *sp*. A decrease in temperature leads to an increase

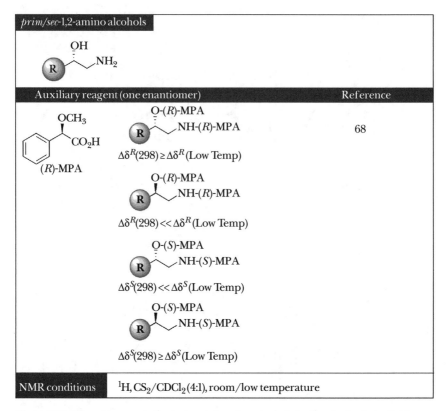

*Figure 5.76.* The assignment of *prim/sec*-1,2-amino alcohols by the single-derivatization method at a glance: CDA, diagnostic shifts ($\Delta\delta^R$, $\Delta\delta^S$), experimental conditions, and references.

in the number of molecules in the most stable conformation [i.e., *ap-sp-gt-II* and *ap-sp-gg-I* in the bis-(*R*)-MPA and bis-(*S*)-MPA amido esters, respectively; for a full description of all the conformations, see Section 5.5.1], and therefore this also increases its contribution to the average NMR spectrum [66, 68]. Conversely, the population of the second most stable conformer diminishes at lower temperatures, and correspondingly, its contribution to the average spectrum also decreases [41, 68].

Thus, considering the bis-(*R*)-MPA derivative of the *prim/sec*-1,2-amino alcohol with the absolute configuration shown in Figure 5.77 (type A) in the most stable conformation [66, 68] (i.e., *ap-sp-gt-II*, left), the phenyl ring of the MPA amide unit shields the MPA ester unit, while in the least stable conformation (i.e., *sp-ap-gt-II*, right), these roles are reversed (i.e., the MPA ester shields the signals of the MPA amide).

A reduction in temperature modifies the equilibrium by increasing the number of molecules in the most stable conformation [68] (i.e., *ap-sp-gt-II*), that is, the number of molecules in which the MPA ester signals are shielded [68]. Naturally, the population of the second conformer (i.e., *sp-ap-gt-II*) decreases in the same ratio, and therefore the number of molecules in which the MPA amide signals are shielded also decreases [68].

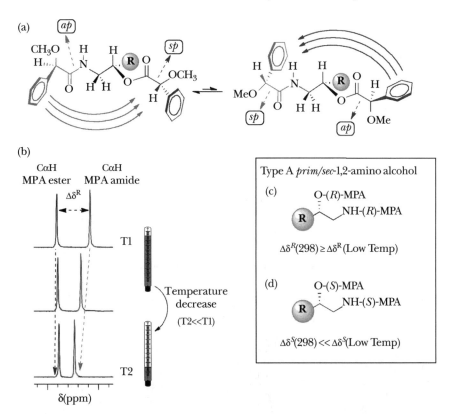

*Figure 5.77.* Simplified procedure for assignment of type A *prim/sec*-1,2-amino alcohols based on $\Delta\delta^R$ and $\Delta\delta^S$ parameters from CαH at different temperatures in bis-MPA esters.

In summary, a drop in temperature is followed by an upfield shift (i.e., more shielding; positive $\Delta\delta^{T1T2}$) of the CαH of the MPA ester [68, 165], and by a downfield shift (i.e., more deshielding; negative $\Delta\delta^{T1T2}$) of the CαH due to the MPA amide unit [58, 165] (Figures 5.77b and c).

The effects transmitted by the MPA amide on the MPA ester are much weaker than those of the MPA ester on the MPA amide, and so the CαH of the MPA ester shifts only slightly with temperature and much less than the CαH signal of the MPA amide.

At room temperature, the separation between the CαH signals from the MPA ester and the MPA amide is equal to or larger than that at a lower temperature [68] [i.e., $\Delta\delta^R(298) \geq \Delta\delta^R$ (low temp); Figures 5.77b and c].

If the bis-(S)-MPA derivative of the same amino alcohol is used instead of the bis-(R)-MPA derivative, examination of the main conformations and shieldings leads to the opposite effect [68] with temperature [i.e., $\Delta\delta^S(298) << \Delta\delta^S$(low temp); Figure 5.77d].

A similar analysis of the bis-(R)-MPA derivatives of the enantiomeric amino alcohol [68] (type B) is presented in Figure 5.78. In the most stable conformer of the bis-(R)-MPA derivative (i.e., *ap-sp-gt-II*, left), the CαH of the MPA amide is

184 ■ The Assignment of the Absolute Configuration by NMR

shielded by the MPA ester, and the CαH of the MPA ester is deshielded by the carbonyl of the MPA ester, whereas in the second most stable conformation (i.e., *sp-ap-gt-II*, right), it is the MPA ester that is shielded by the MPA amide (Figure 5.78a). As the temperature decreases, the CαH signal of the MPA ester is deshielded (i.e., negative $\Delta\delta^{T1T2}$; Figure 5.78b) and that of the MPA amide is shielded (i.e., positive $\Delta\delta^{T1T2}$; Figure 5.78b). The overall effect is a shorter separation between the two CαH signals at room temperature than at a lower temperature [68] [i.e., $\Delta\delta^R(298)$ << $\Delta\delta^R$(low temp); Figure 5.78c].

With the bis-(S)-MPA derivative, the opposite [68] is observed [i.e., $\Delta\delta^R(298) \geq \Delta\delta^R$(low temp); Figure 5.78d].

### 5.5.8. Example 51: Assignment of the Absolute Configuration of (S)-1-Aminopropan-2-ol by Low-Temperature NMR of a Single Derivative

To assign the absolute configuration of a sample of 1-aminopropan-2-ol, the bis-(R)-MPA derivative is prepared (Figure 5.79a), and the corresponding $^1$H-NMR spectra at 298 and 213 K are registered and compared (Figures 5.79b and d),

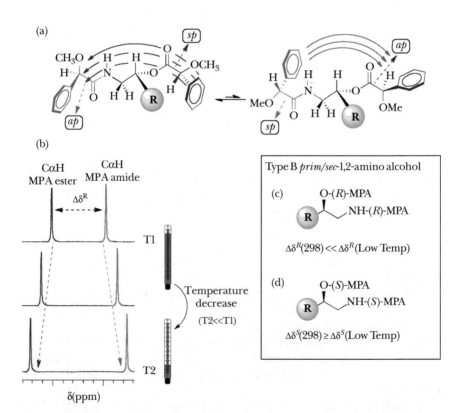

*Figure 5.78.* Simplified procedure for assignment of type B *prim/sec*-1,2-amino alcohols based on $\Delta\delta^R$ and $\Delta\delta^S$ parameters from CαH at different temperatures in bis-MPA esters.

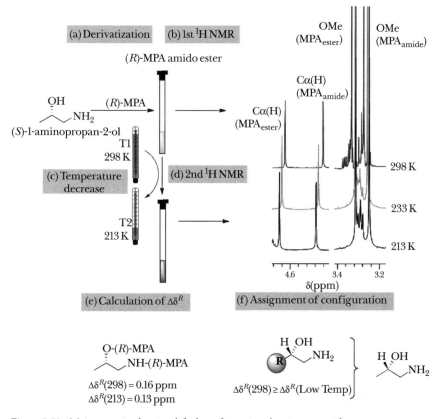

*Figure 5.79.* Main steps in the simplified configurational assignment of (S)-1-aminopropan-2-ol based on CαH diagnostic signals from the bis-(R)-MPA amido ester at different temperatures.

focusing attention on the spectral zone corresponding to the singlets of the CαH of both MPA moieties. The separation between the CαH singlets is calculated at room temperature and at a lower temperature (Figure 5.79e): $\Delta\delta^R(298) = +0.16$ ppm and $\Delta\delta^R(213) = +0.13$ ppm.

As $\Delta\delta^R(298) \geq \Delta\delta^R(\text{low temp})$, the absolute configuration of this amino alcohol is the one shown in Figure 5.77c (type A), namely, (S)-1-aminopropan-2-ol (Figure 5.79f).

The evolution of the $^1$H-NMR spectra can be explained by considering the conformational equilibrium defining the bis-(R)-MPA derivative of (S)-2-aminopropan-1-ol (Figures 5.77a and 5.80). In the most stable conformer (i.e., *ap-sp-gt-II*), the CαH of the MPA ester unit is shielded by the MPA amide and deshielded by the carbonyl of the MPA ester. In the second most stable conformer (i.e., *sp-ap-gt-II*), the CαH of the MPA amide is shielded by the MPA ester.

A drop in temperature produces an increase in the number of molecules in which the CαH of the MPA ester is shielded by the MPA amide and deshielded by the ester carbonyl. Overall, this signal is slightly shifted downfield. The number of

bis-(R)-MPA

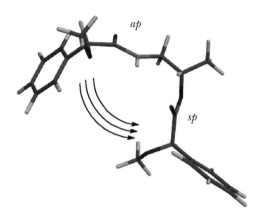

*Figure 5.80.* Main shielding effects caused by MPA phenyl groups in the bis-(R)-MPA amido ester of (S)-1-aminopropan-2-ol.

molecules in which the MPA amide signals are shielded decreases, and the CαH of the MPA amide shifts downfield. Consequently, at room temperature, the separation of the singlets due to the two CαH protons is equal to or greater than it is at a lower temperature [68].

### 5.5.9. Example 52: Assignment of the Absolute Configuration of (R)-1-Aminopropan-2-ol by Low-Temperature NMR of a Single Derivative

A small sample of 2-aminopropan-1-ol is derivatized with (R)-MPA (Figure 5.81a), the corresponding $^1$H-NMR spectra are registered at 298 and 213 K, and the spectra are compared (Figure 5.81b). The separation between the CαH singlets is calculated at room temperature and at a lower temperature (Figure 5.81e). The values obtained are $\Delta\delta^R(298) = +0.34$ ppm and $\Delta\delta^R(213) = +0.51$ ppm.

As $\Delta\delta^R(298) << \Delta\delta^R$(low temp), the absolute configuration of the amino alcohol corresponds to that shown in Figure 5.78c (type B), namely (R)-1-aminopropan-2-ol (Figure 5.81f).

This evolution of the spectra can be explained by considering the conformational equilibrium shown in Figure 5.78a. In the most stable conformer (i.e., *ap-sp-gt-II*), the CαH of the MPA amide is shielded by the MPA ester. In the second most stable conformer (i.e., *sp-ap-gt-II*), the CαH of the MPA ester is shielded by the MPA amide.

A reduction in temperature produces an increase in the number of molecules in which the CαH of the MPA amide is shielded, and a decrease in the number of molecules in which the MPA ester signals are shielded. Consequently, the separation of the singlets due to the two CαH protons is lower at room temperature than it is at a lower temperature.

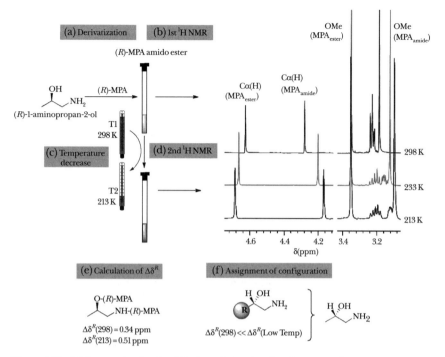

*Figure 5.81.* Main steps in the simplified configurational assignment of (R)-1-aminopropan-2-ol based on CαH diagnostic signals from the bis-(R)-MPA amido ester at different temperatures.

## 5.5.10. Summary

As in the case of *sec/prim*-1,2-aminoalcohols, the absolute configuration of *prim/sec*-1,2-amino alcohols can be determined by either double- or single-derivatization methods [13, 66–68, 165] using MPA as the CDA and $^1$H NMR. The double-derivatization procedure [66, 67] requires the comparison of the $^1$H-NMR spectra of both bis-MPA derivatives. Once again, two sets of signals [66–68] can be used from either the R and $CH_2(1')$ hydrogens or the OMe and CαH hydrogens of the MPA auxiliaries. In both cases, their $\Delta\delta^{RS}$ values must be compared for the assignment.

The OMe and CαH signals can also be used in a different way (as in the case of *sec/prim*-1,2-amino alcohols), that is, by looking at the separation between the CαH and OMe signals in the bis-(R)-MPA ($\Delta\delta^R$) and in the bis-(S)-MPA derivative ($\Delta\delta^R$).

In the single-derivatization method [68], the evolution with temperature of the signals due to the CαH protons of the two MPA units (MPA amide and MPA ester) is analyzed in only one bis-MPA derivative. The separation between those two singlets at room temperature and at a lower temperature [e.g., $\Delta\delta^R(298)$ and $\Delta\delta^R$(low temp)] is used for the assignment.

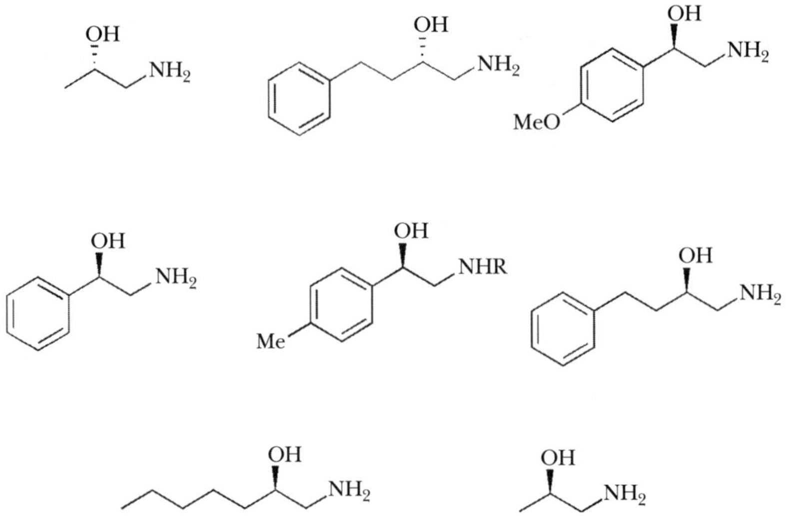

*Figure 5.82.* Selection of *prim/sec*-1,2-amino alcohols with known absolute configuration that have been used to validate the procedures described in this chapter.

No procedure for assignment of this class of compounds has been demonstrated based on $^{13}$C NMR [72].

Figure 5.82 shows a selection of *prim/sec*-1,2-amino alcohols that have been used to validate the procedures described in this chapter. Other examples of applications of this methodology can be found in the literature [217–220].

## 5.6. PRIM/SEC/SEC-1,2,3-TRIOLS

*Prim/sec/sec*-1,2,3-triols constitute the most complex functional array whose absolute configuration can be determined by direct derivatization of the three hydroxyl groups and NMR comparison of the resulting tris-(*R*)- and tris-(*S*)-MPA ester derivatives [13, 69, 70] (Figure 5.83).

### 5.6.1. Double-Derivatization Method: MPA

The complexity of this system arises from the presence of three MPA units in the ester derivative and the difficulty of finding a correlation between the stereochemistry of the two asymmetric carbons and the chemical shifts that are determined by the combination of shielding effects due to three phenyl rings [69, 70].

Nevertheless, having in mind the conformational characteristics of the MPA derivatives of *prim/sec*- and *sec/sec*-1,2-diols [59–61, 63–65] discussed in Sections 5.1 and 5.3, it is not too complex to carry out conformational studies of the tris-(*R*)- and tris-(*S*)-MPA ester derivatives of *prim/sec/sec*-1,2,3-triols [69, 60] (Figure 5.84b). Those results allow us to figure out how each of the three MPA units and their phenyl rings are orientated relative to the rest of the molecule and therefore

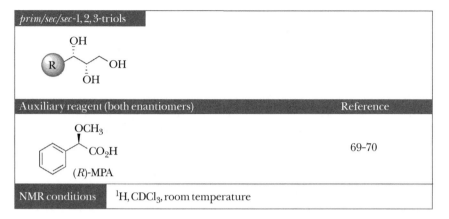

*Figure 5.83.* The assignment of *prim/sec/sec*-1,2,3-triols at a glance: CDA, experimental conditions, and references.

(a) Diastereoisomers of *prim/sec/sec*-1, 2, 3-triols

syn Type A    syn Type B    anti Type C    anti Type D

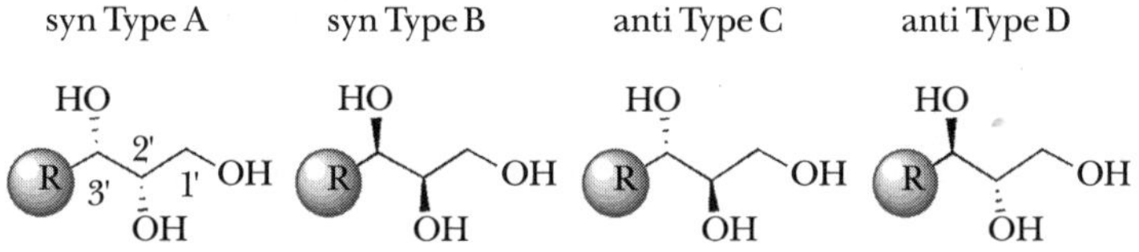

(b) Generation of main conformers

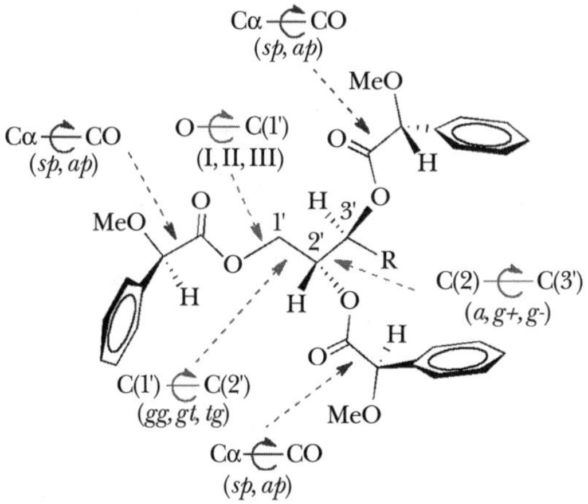

*Figure 5.84.* Classification of *prim/sec/sec*-1,2,3-triols according to structural types and main bonds involved in conformational equilibria.

190 ■ The Assignment of the Absolute Configuration by NMR

to predict which protons are located under their shielding cones in each derivative of the four possible stereoisomers (Figure 5.84a).

The conformational study [69, 70] indicates that, in *prim/sec/sec*-1,2,3-triols, the methine protons [i.e., H(2′) and H(3′); Figure 5.84] are the signals of diagnostic value because their shifts correlate with the stereochemistry.

Figure 5.85 shows the result of the combined effect of the three MPA units on the chemical shifts of H(2′) and H(3′) in the tris-(R)- and the tris-(S)-MPA derivatives of the four stereoisomers. It is obvious that the tris-(R)- and the tris-(S)-MPA esters of each stereoisomer will present different chemical shifts for the diagnostic signals and therefore distinct $\Delta\delta^{RS}$ values.

Unfortunately, the resulting $\Delta\delta^{RS}$ signs from H(2′) and H(3′) are not sufficient to distinguish among the stereoisomers because H(2′) is under comparable shielding

(a) Shielded protons in the tris-MPA esters of type A *syn*-triols

(b) Shielded protons in the tris-MPA esters of type B *syn*-triols

(c) Shielded protons in the tris-MPA esters of type C *anti*-triols

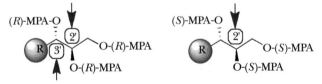

(d) Shielded protons in the tris-MPA esters of type D *anti*-triols

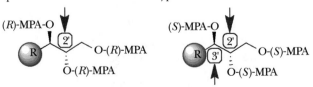

*Figure 5.85.* Shielding effects on H(2′) and H(3′) in the tris-MPA esters of *prim/sec/sec*-1,2,3-triols.

effects in both of the tris-MPA esters of type C *anti*-triols and in both tris-MPA esters of type D *anti*-triols; its $\Delta\delta^{RS}$ signs in those cases are unpredictable.

Thus, the distribution of shielding effects and resulting $\Delta\delta^{RS}$ signs for type A *syn*-triols [$\Delta\delta^{RS}H(2') > 0$ and $\Delta\delta^{RS}H(3') > 0$], for type B *syn*-triols [$\Delta\delta^{RS}H(2') < 0$ and $\Delta\delta^{RS}H(3') < 0$], for type C *anti*-triols [unpredictable$\Delta\delta^{RS}H(2')$ and $\Delta\delta^{RS}H(3') < 0$], and for type D *anti*-triols [unpredictable$\Delta\delta^{RS}H(2')$ and $\Delta\delta^{RS}H(3') > 0$] do not allow distinguishing type A from type D or type B from type C.

This uncertainty can be solved by noting the special characteristics of the effect of the MPA unit located at the primary alcohol. This unit acts on H(2') and H(3') without modifying the $\Delta\delta^{RS}$ signs produced by the other two MPAs, and it contributes to the $\Delta\delta^{RS}$ signs in the same sense as the MPA units on secondary alcohols; however, the effects on one of those protons is much more intense than it is on the other. As a result, the relative intensity of $\Delta\delta^{RS}$, measured as absolute value, on H(2') and H(3'), can be used as an experimental parameter that correlates with the stereochemistry. In fact, type A *syn*- and type B *syn*-triols share a very small absolute value; it is close to 0, in the range of 0.00–0.06 ppm for $|\Delta(\Delta\delta^{RS})|$ [defined as $|\Delta(\Delta\delta^{RS})|=|\Delta\delta^{RS}H(2') - \Delta\delta^{RS}H(3')|$], while the type C and type D *anti*-triols share a larger value that is about 0.16–0.53 ppm for the same difference.

Thus, using the combination of the positive and negative $\Delta\delta^{RS}$ signs from H(3') [i.e., $\Delta\delta^{RS}H(3')$] and the absolute values of $|\Delta(\Delta\delta^{RS})|$, it is possible to distinguish between the four stereoisomers of a *prim/sec/sec*-1,2,3-triol.

Figure 5.86 summarizes the correlation between the absolute stereochemistry of the triols and the NMR spectra [69, 70] of the tris-(*R*)- and the tris-(*S*)-MPA

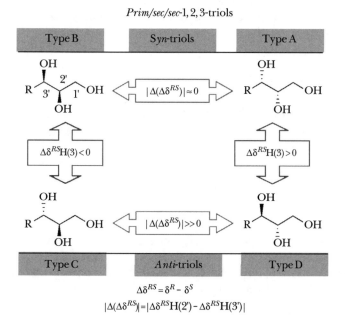

*Figure 5.86.* Assignment of the absolute configuration of *prim/sec/sec*-1,2,3-triols based on the combined used of $\Delta\delta^{RS}H(3')$ and $|\Delta(\Delta\delta^{RS})|$ parameters from tris-MPA esters.

ester derivatives, using the methine protons H(2') and H(3') as diagnostic signals and the $\Delta\delta^{RS}$ sign of H(3') and the $|\Delta(\Delta\delta^{RS})|$ value of H(2') and H(3') as parameters for the assignment.

From a practical point of view, the procedure for assignment is the following:

1) The tris-(R)- and the tris-(S)-MPA esters are prepared.
2) Their ¹H-NMR spectra are registered, paying attention to the signals from the methine protons, H(2') and H(3').
3) The corresponding $\Delta\delta^{RS}$ signs and values are calculated, focusing on the $\Delta\delta^{RS}$H(3') sign. According to Figure 5.86, a positive $\Delta\delta^{RS}$ sign for H(3') indicates that the triol has either a type A *syn-* or a type D *anti-*structure, while a negative $\Delta\delta^{RS}$ sign for H(3') indicates either a type B *syn-* or a type C *anti-*triol structure.
4) To distinguish between those possibilities, the quantitative difference $|\Delta(\Delta\delta^{RS})|$ (absolute value) is calculated as the $\Delta\delta^{RS}$ value for H(2') minus that for H(3'). When this difference is very small, close to zero (we observe values in the range of 0.00–0.06 ppm in our test compounds), the triol structure is either type A *syn* or type B *syn*. If the difference is large, with values in the range of 0.16–0.53 ppm (obtained in a series of test compounds), the triol has a type C *anti* or type D *anti* structure.
5) Combining the two parameters $\Delta\delta^{RS}$H(3') and $|\Delta(\Delta\delta^{RS})|$ allows us to accurately distinguish any one of the four stereoisomers.

Those correlations between the stereochemistry and the spectra have been tested experimentally with a series of *prim/sec/sec*-triols of known absolute configuration derivatized as tris-MPA esters [69, 70]. A complete agreement between the ¹H-NMR prediction and the absolute configuration was obtained in all cases.

### 5.6.2. Example 53: Assignment of the Absolute Configuration of Hexane-1,2,3-Triol (*Syn*)

Figure 5.87 illustrates the above procedure with the detailed assignment of configuration of a sample constituted by a pure enantiomer of hexane-1,2,3-triol derivatized as tris-(R)- and tris-(S)-MPA ester (Figure 5.87a). The two ¹H-NMR spectra are registered in deuterated chloroform at room temperature, the signals for the methine protons [H(2') and H(3')] are assigned (Figure 5.87b), and the $\Delta\delta^{RS}$ values and signs for H(2') and H(3') are calculated (+0.12 and +0.16, respectively), as well as their $|\Delta(\Delta\delta^{RS})|$ difference (as an absolute value, $|\Delta(\Delta\delta^{RS})| = |\Delta\delta^{RS}H(2') - \Delta\delta^{RS}(3')| = |-0.04| = 0.04$ ppm) (Figure 5.87c).

Comparison of those parameters [positive $\Delta\delta^{RS}H(3')$; small $|\Delta(\Delta\delta^{RS})|$] with the signs and values in Figure 5.86 show that this coincides with a type A *syn*-triol, and therefore the compound has the absolute configuration shown in Figure 5.87d [i.e., (2S, 3S)-hexane-1,2,3-triol].

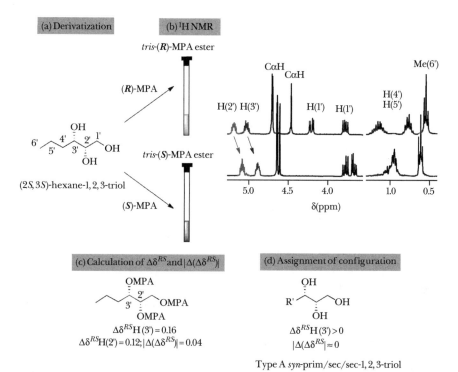

Figure 5.87. Assignment of the absolute configuration of *prim/sec/sec*-1,2,3-triols based on the combined used of $\Delta\delta^{RS}H(3')$ and $|\Delta(\Delta\delta^{RS})|$ parameters from tris-MPA esters.

### 5.6.3. Example 54: Assignment of the Absolute Configuration of Hexane-1,2,3-Triol (*Anti*)

A stereoisomer of hexane-1,2,3-triol is derivatized as tris-(*R*)- and tris-(*S*)-MPA ester (Figure 5.88a), the corresponding ¹H-NMR spectra are registered (Figure 5.88b), the signals for the methine protons [H(2′) and H(3′)] are assigned, and the values of $\Delta\delta^{RS}H(2')$, $\Delta\delta^{RS}H(3')$, and $|\Delta(\Delta\delta^{RS})|$ are calculated to be −0.04, +0.41, and |−0.045| = 0.45 ppm, respectively (Figure 5.88c).

Comparison of those parameters [positive $\Delta\delta^{rs}H(3')$; large $|\Delta(\Delta\delta^{RS})|$] with the signs and values in Figure 5.86 shows that they coincide with those of a type D *anti*-triol, and therefore the compound has the absolute configuration shown in 5.88d [i.e., (2*S*, 3*R*)-hexane-1,2,3-triol].

### 5.6.4. Summary

The absolute configuration of *prim/sec/sec*-1,2,3-triols can be determined by double derivatization with (*R*)- and (*S*)-MPA and comparison of the ¹H-NMR spectra of the tris-MPA esters using the methine (2′)/(3′) hydrogens as diagnostic signals [13, 69–70].

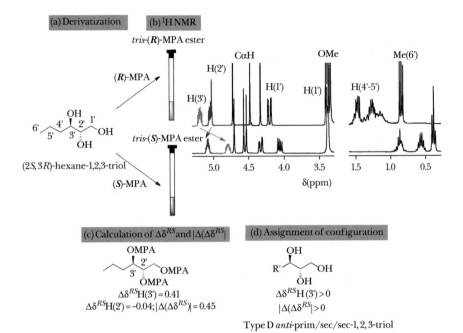

*Figure 5.88.* Main steps in the assignment of (2S, 3R)-hexane-1,2,3-triol based on $\Delta\delta^{RS}H(3')$ and $|\Delta(\Delta\delta^{RS})|$ parameters from the tris-MPA esters.

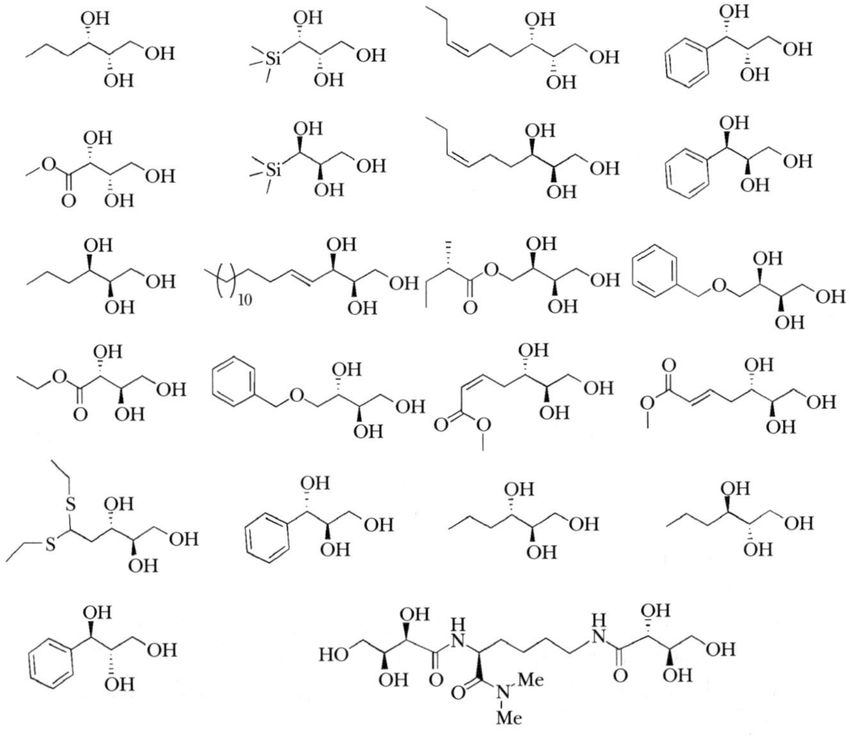

*Figure 5.89.* Selection of *prim/sec/sec*-1,2,3-triols with known absolute configuration that have been used to validate the procedure described in this chapter.

Experimental and theoretical data show that there is a correlation between the stereochemistry and the $^1$H-NMR spectra based on the $\Delta\delta^{RS}$ sign of H(3′) and the $|\Delta(\Delta\delta^{RS})|$ difference (absolute value) between the $\Delta\delta^{RS}$ of H(2′) and H(3′). Using these two parameters, the four stereoisomers of a *prim/sec/sec*-1,2,3-triol can be easily distinguished.

To date, no procedures based on the use of $^{13}$C NMR nor those by single derivatization have been described for triols [72].

Figure 5.89 shows a selection of *prim/sec/sec*-1,2,3-triols [69–70] of known absolute configuration that have been used to validate the procedures for assignment presented in this chapter.

# 6 Exercises

This chapter contains fifty exercises in the use of the methods described in the previous pages to the assignment of the absolute configuration of different substrates. They are organized according to the functional group of the substrate and follow the same order as in the previous chapters: alcohols; cyanohydrins; thiols; amines; carboxylic acids; 1,n-diols; 1,2-amino alcohols; and 1,2,3-triols. For each class of functional groups, there are exercises related to all the procedures available for assignment, such as double and single derivatization, and with different CDAs, when this is relevant.

The derivatives and their spectra are actual data obtained in our laboratory; they are not simulated. Therefore, in a few cases, the spectra is not of high quality due to a small sample size or impurities that affect the result.

In the spectra provided in the exercises below, in most cases, we have labeled the signals relevant for assignment; however, when the structure of the substrate and the spectra are simple, no signal identification is provided.

At the end of this chapter, we provide a list of references to the literature in which the solutions can be found.

**Exercise 1.** The (R)- and (S)-9-AMA ester derivatives of a pure enantiomer of the terpene isopinocampheol were prepared, and their $^1$H NMR were registered. They are shown below. Can you deduce the correct absolute stereochemistry of the secondary alcohol from these spectra?

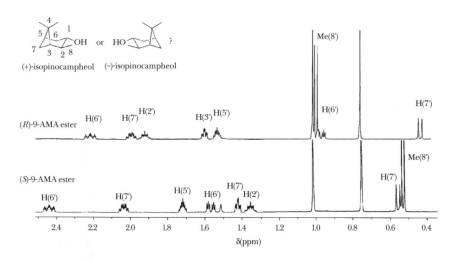

198 ■ The Assignment of the Absolute Configuration by NMR

**Exercise 2.** A sample of a pure isomer of borneol was derivatized with (R)- and (S)-MPA. The ¹H-NMR spectra of the corresponding esters are shown in the figure below. Deduce the absolute stereochemistry, and tell which one of the two enantiomers shown is the one that matches the sample provided.

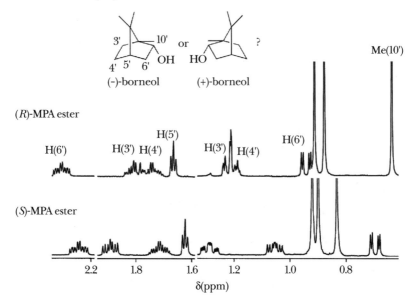

**Exercise 3.** An enantiomer of the secondary alcohol shown below was obtained by stereospecific reduction. A small amount was derivatized with a 2:1 mixture of (R)- and (S)-1-NMA. The figure shows the ¹H-NMR spectrum of the resulting mixture of ester derivatives. From those data, can you assign the absolute stereochemistry of that secondary alcohol?

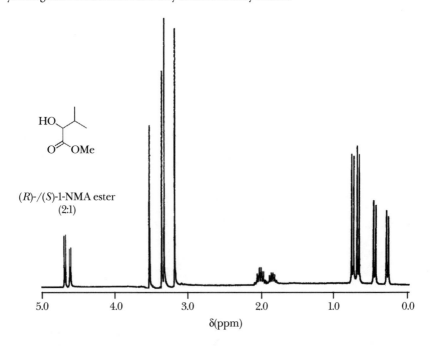

**Exercise 4.** A pure enantiomer of menthol was obtained from a natural source. In order to know its absolute stereochemistry, a small amount was derivatized with a (3:1) mixture of (R)- and (S)-MPA. The $^{13}$C-NMR spectrum of the mixture of ester derivatives is shown below. Can you assign the absolute stereochemistry and identify the particular enantiomer of this sample?

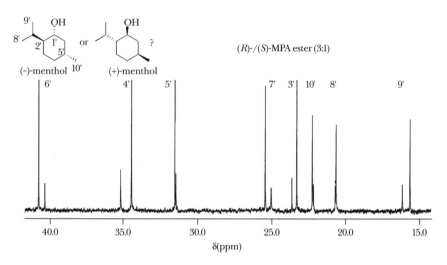

**Exercise 5.** The (R)- and (S)-9-AMA esters of a sample of (?)-1,2-isopropylidene-glycerol were prepared, and their $^1$H-NMR spectra were recorded. Does the sample correspond to the (R)- or to the (S)-enantiomer?

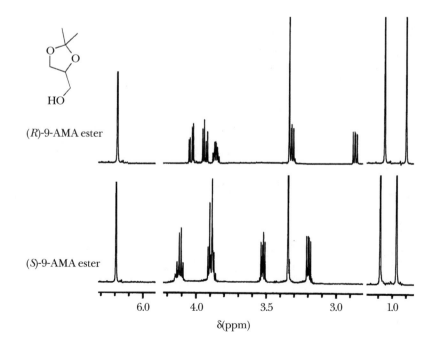

**Exercise 6.** In order to assign the absolute configuration of an enantiomer of (?)-methyl 3-hydroxy-2-methylpropanoate, the corresponding (R)- and (S)-9-AMA esters were synthesized, and their ¹H-NMR spectra were compared (see figure below). Determine the stereochemistry of the enantiomer in question.

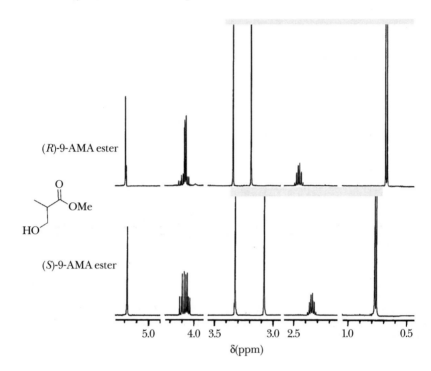

**Exercise 7.** The figure below shows the ¹H-NMR spectra of the (R)- and the (S)-MPA ester derivatives of one enantiomer of the ketone cyanohydrin that is shown. Can you identify the enantiomer using these data?

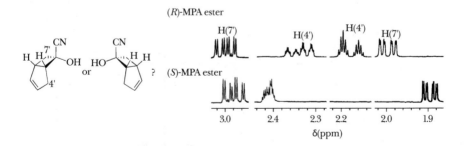

**Exercise 8.** The absolute configuration of the cyanohydrin shown in the figure below was determined by comparison of the $^1$H-NMR spectra of the corresponding (R)- and (S)-MPA esters. What is the absolute stereochemistry of the cyanohydrin?

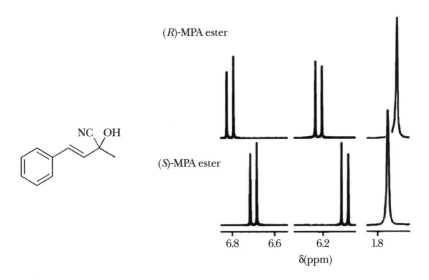

**Exercise 9.** The (R)- and (S)-2-NTBA thioester derivatives of the chiral thiol shown in the figure below were prepared, and their $^1$H-NMR spectra were recorded. Can you determine the absolute configuration of the C(1') chiral center of the thiol?

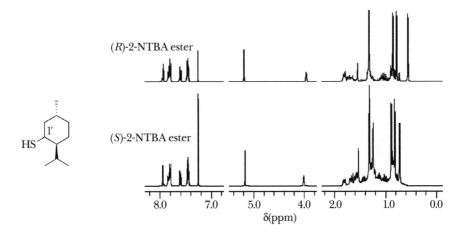

**Exercise 10.** A sample of an enantiomer of (?)-ethyl 2-mercaptopropanoate was derivatized with (R)- and (S)-MPA. The ¹H-NMR spectra of the thioesters are shown in the figure. Use those data to propose the absolute configuration of this thiol.

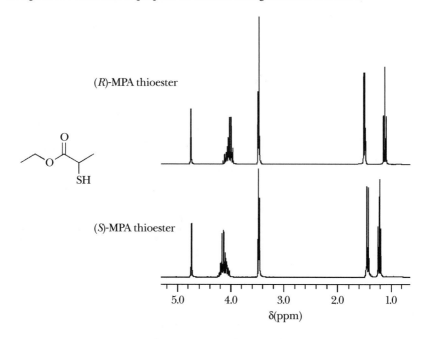

**Exercise 11.** In order to identify an unknown stereoisomer of bornylamine, the (R)- and (S)-BPG amides were prepared, and their ¹H-NMR spectra were registered. Can you propose the absolute configuration from that information?

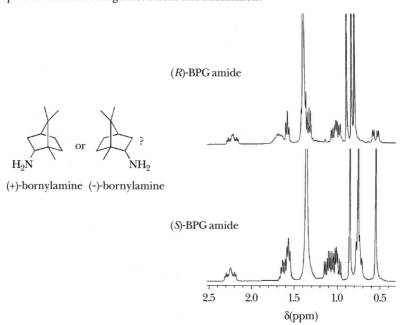

Exercises ■ 203

**Exercise 12.** The (R)- and (S)-MPA amides of a phenylalanine methyl ester of unknown stereochemistry were prepared, and their ¹H-NMR spectra were registered. They are shown below. Can you determine the absolute configuration of the amine?

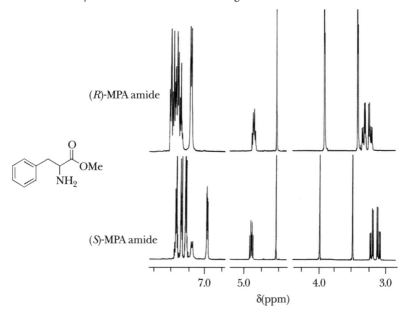

**Exercise 13.** To determine the absolute configuration of a chiral (?)-butan-2-amine, a small amount was derivatized with a (2:1) mixture of (R)- and (S)-MPA. The ¹H-NMR spectrum of the resulting mixture of diastereoisomers is shown below. What is the configuration of the amine?

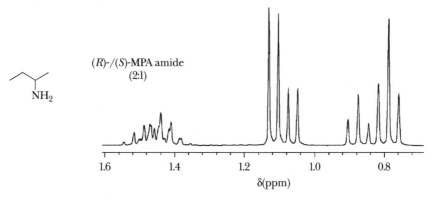

**Exercise 14.** The ¹H-NMR spectra of the (R)- and (S)-BPG amide derivatives of a chiral leucine methyl ester are shown below. Indicate which is the absolute configuration of the amino acid ester.

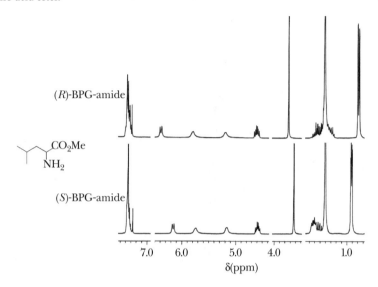

**Exercise 15.** The (R)- and (S)-9-AHA esters of a sample of (?)-2-methylbutanoic acid of unknown configuration were prepared, and their ¹H-NMR spectra were recorded. Does the sample correspond to the (R)- or the (S)-enantiomer?

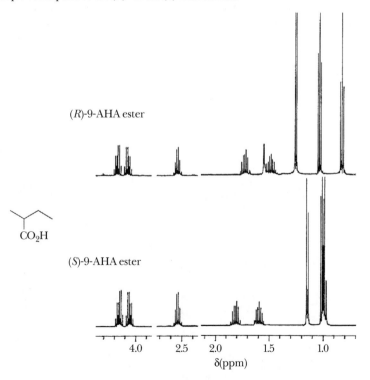

**Exercise 16.** A very small sample of one enantiomer of (?)-butan-2-ol was obtained in the course of a synthesis. To determine its absolute configuration, the ¹H-NMR spectra of its (R)-MPA ester were recorded at different temperatures (300 K and 203 K). Can you assign the absolute configuration of that alcohol from these spectra?

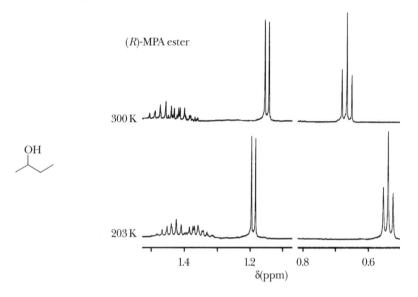

**Exercise 17.** A sample of (?)-pentan-2-ol of unknown configuration was derivatized with (S)-MPA, and its ¹H-NMR spectra were recorded at different temperatures (300 K and 203 K). Propose the absolute configuration of the alcohol.

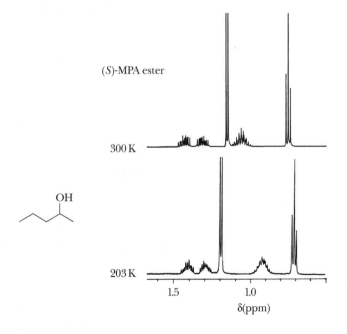

**Exercise 18.** The absolute configuration of the secondary alcohol shown in the figure was determined by analysis of the evolution of the ¹H-NMR spectra of its (S)-MPA ester derivative upon addition of Ba(ClO₄)₂. What is the absolute configuration of the alcohol?

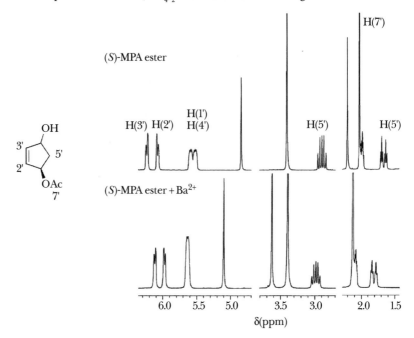

**Exercise 19.** The two enantiomers of menthol are present in nature. A sample of (?)-menthol of unknown configuration was derivatized with (S)-MPA. The figure shows its ¹H-NMR spectra in MeCN-d₃ before and after addition of Ba(ClO₄)₂. Can you assign the absolute configuration of that enantiomer of menthol from comparison of these spectra?

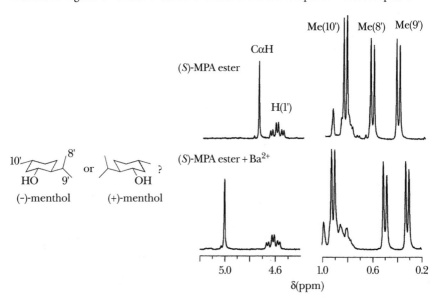

Exercises ■ 207

**Exercise 20.** The (R)-MPA ester of a diacetone glucose of unknown stereochemistry was prepared, and its ¹H-NMR spectra were recorded before and after addition of Ba(ClO$_4$)$_2$. What is the stereochemistry at C(3′)?

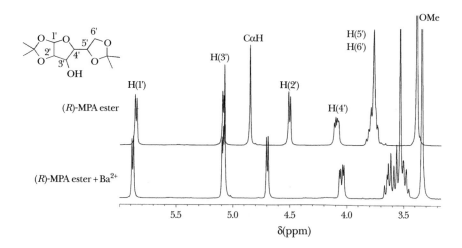

**Exercise 21.** In order to identify the enantiomer of an isopinocampheylamine sample, it was derivatized with (R)-MPA, and the ¹H-NMR spectra of the (R)-MPA amide were recorded in the presence and absence of Ba(ClO$_4$)$_2$. What is the correct structure of the compound, (−)- or (+)-isopinocampheylamine?

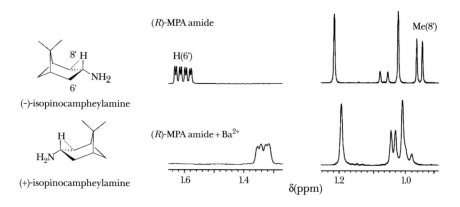

**Exercise 22.** The figure shows the evolution of the ¹H-NMR spectra of the (R)-MPA amide of a bornylamine of unknown stereochemistry after addition of Ba(ClO$_4$)$_2$. Can you assign the absolute configuration of this amine?

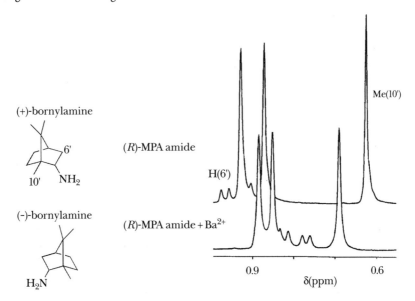

**Exercise 23.** The two enantiomers of 3,3-dimethylbutan-2-ol were separated by chiral high-performance liquid chromatography (HPLC) and labeled as A and B. In order to unravel their absolute configurations, the two samples were separately esterified with (R)-9-AMA. The ¹H-NMR spectra of alcohols A and B and those of their (R)-9-AMA esters are shown in the figure below. Can you determine the absolute configurations of A and B?

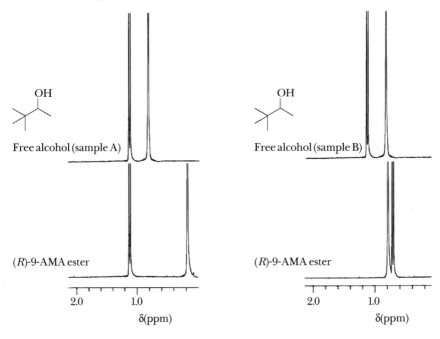

**Exercise 24.** The absolute configuration of the marine metabolite shown in the figure below was assigned by comparison of the ¹H-NMR spectra of the alcohol with that of its (S)-9-AMA ester. Can you assign the configuration at C(8′) of this natural product?

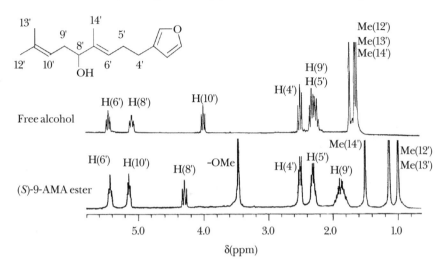

**Exercise 25.** The ¹H-NMR spectra of the bis-(R)- and the bis-(S)-MPA ester derivatives of a pure stereoisomer of 2,3-dihydroxyhexyl acetate are shown below. Propose its absolute configuration.

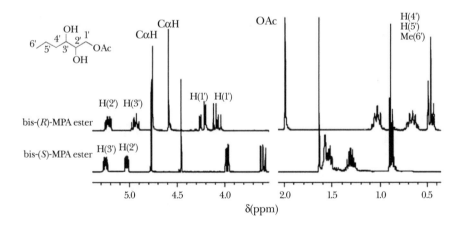

**Exercise 26.** Another sample of 2,3-dihydroxyhexyl acetate has been obtained, but its absolute configuration is unknown. Suggest the stereochemistry of this diol from the $^1$H-NMR spectra of its bis-(R)- and bis-(S)-MPA esters.

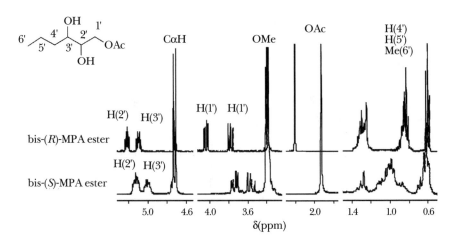

**Exercise 27.** The absolute configuration of *trans*-muriosolinone (an acetogenin) has been assigned by comparison of the $^1$H-NMR spectra of the bis-MTPA esters. The corresponding $\Delta\delta^{SR}$ (ppm) values and signs are shown below. Can you determine the stereochemistry at C(15′) and at C(20′)?

R = MTPA; R = H, *trans*-muriosolinone

**Exercise 28.** To assist the assignment of the absolute configuration of the natural product rolliniastatin-2, the bis-(R)- and bis-(S)-2-NMA ester derivatives were prepared. Determine the stereochemistry at C(24′) and at C(15′) using the $\Delta\delta^{RS}$ data shown in the figure below.

R = 2-NMA; R = H, rolliniastatin-2

**Exercise 29.** The absolute configuration of the diol shown in the figure below is unknown, and only a very few milligrams of the compound are available for analysis. Can you propose its absolute configuration by observing the evolution with temperature of the $^1$H-NMR spectra of its bis-(R)-MPA ester?

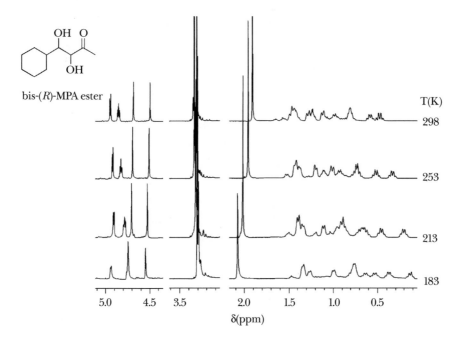

**Exercise 30.** Four samples (A–D) share a common heptano-2,3-diol structure, but they have different stereochemistry. They were derivatized with an excess of (R)-MPA, and the spectra of the bis-(R)-ester derivatives were recorded at room temperature and at a lower temperature (183–298 K). Can you identify the structure of the isomer present in each sample?

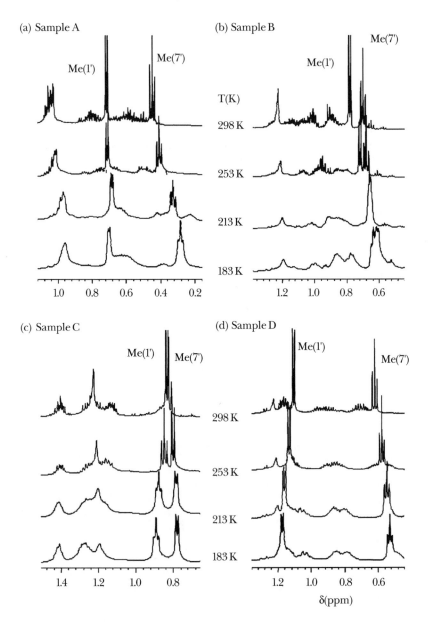

**Exercise 31.** In order to disclose the absolute configuration of the *sec/sec*-amino alcohol shown in the figure, its bis-MPA derivatives were prepared and submitted to ¹H-NMR spectroscopy. Using the spectra, propose the absolute configuration of the amino alcohol.

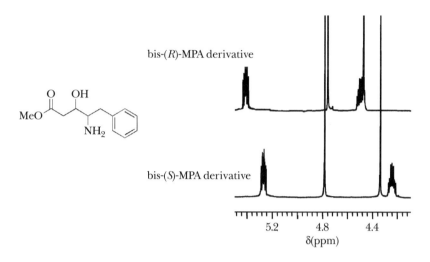

**Exercise 32.** The figure below shows the ¹H-NMR spectra of the two bis-MPA amido esters of a stereoisomer of 2-aminopentan-3-ol. Assign its absolute configuration.

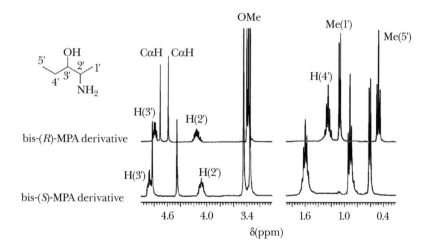

**Exercise 33.** The figure shows the evolution with temperature (183–300 K) of the ¹H-NMR spectra of the bis-(R)-MPA ester of a pure enantiomer of 1-phenylethane-1,2-diol. Can you assign its absolute configuration from these data?

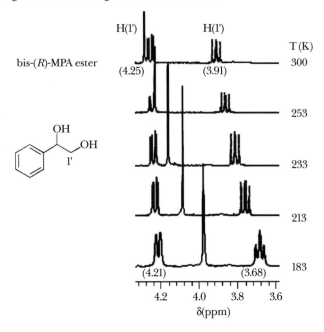

**Exercise 34.** A 3,3-difluoroheptane-1,2-diol of unknown stereochemistry was derivatized with (R)-MPA, and the ¹H-NMR spectra of the bis-(R)-MPA ester were registered at different temperatures (183–300 K). The signals due to the methylene (1′) protons [i.e., $CH_2(1′)$], are shown in the figure below. What is the absolute configuration of the diol?

**Exercise 35.** To determine the absolute configuration of the diol shown in the figure below, its bis-(S)-MPA ester was prepared, and the corresponding ¹H-NMR spectra were recorded at different temperatures (183–300 K). The signals of the methylene (1′) protons are shown in the figure below. Suggest the absolute configuration of the diol.

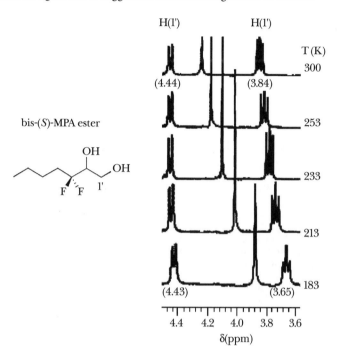

**Exercise 36.** A sample of the diol shown in the figure below was derivatized with (R)- and (S)-9-AMA, and the ¹H-NMR spectra of the resulting bis-9-AMA esters were registered. The signals from the diastereotopic methylene protons [CH$_2$(1′)] are also shown. Assign the absolute configuration at C(2′).

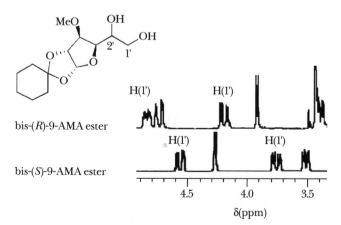

**Exercise 37.** The two bis-9-AMA esters of the diol shown in the figure below were prepared, and the configuration was deduced from the corresponding ¹H-NMR spectra. Which is its absolute configuration?

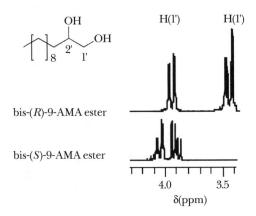

**Exercise 38.** The ¹H-NMR spectra of the two bis-MPA amido esters of an enantiomer of 2-amino-4-(methylthio)butan-1-ol are shown in the figure below. What is the absolute configuration of this amino alcohol?

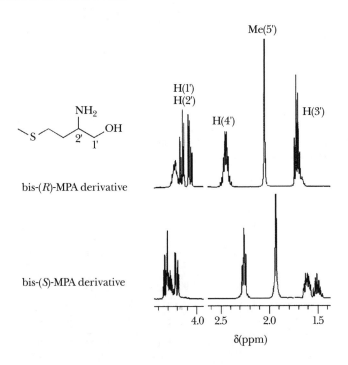

**Exercise 39.** An amino alcohol sample has the structure shown below, but its absolute configuration is unknown. In order to disclose the configuration, the sample was derivatized with (R)- and (S)-MPA. The figure shows partial ¹H-NMR spectra containing the signals due to the CαH and OMe at the auxiliaries. Can you deduce the absolute configuration of the amino alcohol?

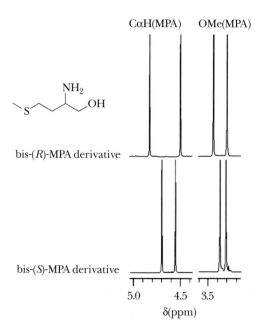

**Exercise 40.** The absolute configuration of a pure stereoisomer of 2-amino-3-methylpentan-1-ol was assigned by ¹H-NMR of the bis-MPA derivatives. Can you determine the absolute configuration at C(2')?

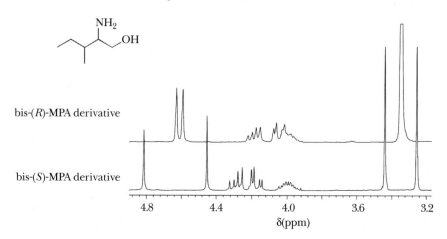

218 ■ The Assignment of the Absolute Configuration by NMR

**Exercise 41.** The figure below shows partial ¹H-NMR spectra of the bis-MPA derivatives of an enantiomer of 2-aminobutan-1-ol. What is the absolute configuration of the amino alcohol?

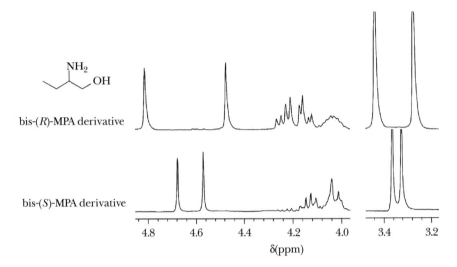

**Exercise 42.** To assign the configuration of a small sample of 2-aminobutan-1-ol of unknown stereochemistry, its bis-(S)-MPA amido ester was prepared, and its ¹H-NMR spectra were recorded at different temperatures (183–298 K). Use the evolution of the spectra with temperature to propose the absolute configuration of the amino alcohol.

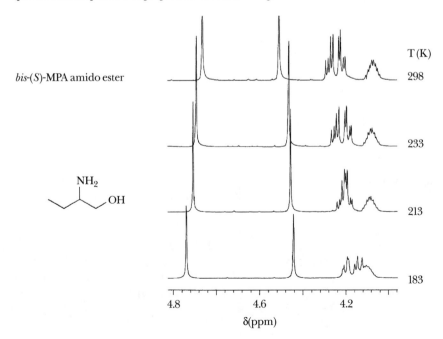

**Exercise 43.** The stereochemistry of a sample of 2-amino-4-methylpentan-1-ol was deduced from the evolution with decreasing temperature of the ¹H-NMR spectra of its bis-(S)-MPA derivative. Determine the configuration of the amino alcohol.

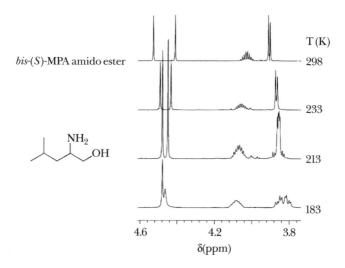

**Exercise 44.** An optically active sample of 1-amino-4-phenylbutan-2-ol of unknown stereochemistry was derivatized with MPA. The figure below shows the NMR spectra of the resulting bis-(R)- and bis-(S)-MPA derivatives. Using this data, assign the absolute configuration of the amino alcohol.

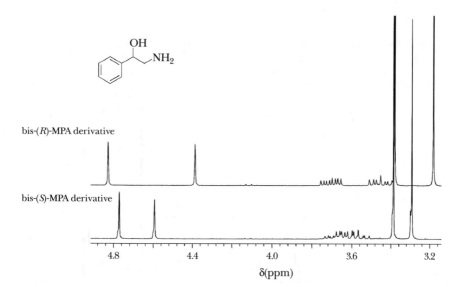

**Exercise 45.** In order to know the absolute configuration of an enantiomer of 1-amino-4-phenylbutan-2-ol, the NMR spectra of the bis-(R)- and bis-(S)-MPA derivatives were compared and analyzed. Can you assign the configuration?

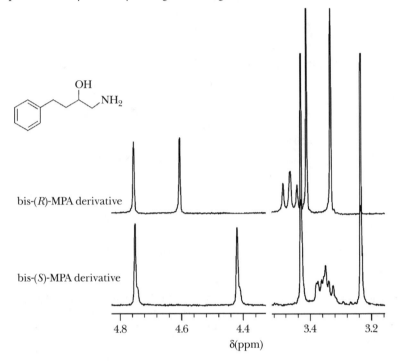

**Exercise 46.** To determine the absolute configuration of the amino alcohol shown in the figure below, the NMR spectra of the bis-(R)-MPA derivative were examined at different temperatures. On the basis of the evolution observed, suggest the absolute configuration of the amino alcohol.

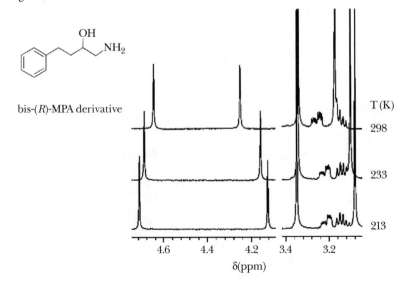

Exercises ■ 221

**Exercise 47.** The amino alcohol shown below, 1-aminoheptan-2-ol, was obtained in chiral form. In order to know its stereochemistry, it was derivatized with (R)-MPA. The spectra of the bis-(R)-MPA amido ester were recorded at different temperatures (183–298 K), and these are shown below. Can you deduce the configuration of the amino alcohol?

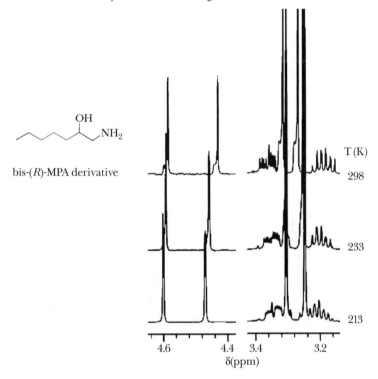

**Exercise 48.** To assign the absolute configuration of the *prim/sec/sec*-1,2,3-triol shown in the figure below, 1-(trimethylsilyl)propane-1,2,3-triol, its tris-MPA esters were prepared and submitted to $^1$H-NMR spectroscopy. Can you assign the absolute configuration of the triol from these partial spectra?

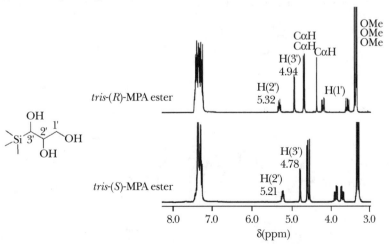

**Exercise 49.** The stereochemistry of the triol shown in the figure below is unknown. Can you use the partial ¹H-NMR spectra of the corresponding tris-MPA esters to assign the configuration at the two asymmetric carbons?

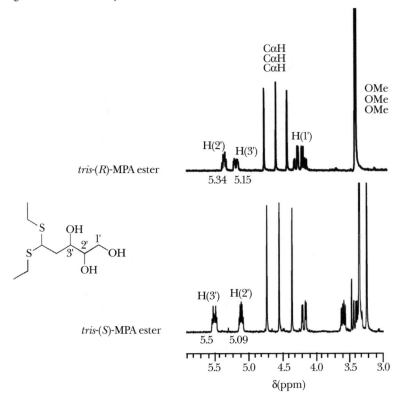

**Exercise 50.** The structure of the compound shown below is formed by two different prim/sec/sec-1,2,3-triol moieties of unknown configuration, and these are connected by a long diamino linker to form a single stereoisomer. The compound was derivatized with an excess of (R)- and (S)-MPA, and the ¹H-NMR spectra of the resulting *hexakis*-MPA esters are presented below. Using this information, propose the stereochemistry at C(2′)/C(3′) and at C(2″)/C(3″).

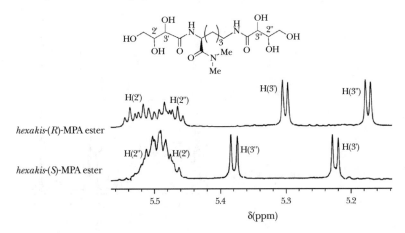

## ■ SOLUTIONS

1. Reference 14, compound 48.13.
2. Reference 14, compound 49.3.
3. Reference 14, enantiomer of compound 48.10.
4. Reference 14, compound 49.4.
5. Reference 45, compound 9.
6. Reference 45, compound 10.
7. Reference 48, compound 2.
8. Reference 48, compound 9.
9. Reference 50, compound 7.
10. Reference 50, enantiomer of compound 8.
11. Reference 54, compound 10.
12. Reference 52, compound 8.
13. Reference 56, enantiomer of compound 5.
14. Reference 54, compound 13.
15. Reference 57, compound 6.
16. Reference 41, compound 4.
17. Reference 41, compound 5.
18. (1$R$, 4$S$)-4-hydoxycyclopent-2-en-1-yl acetate.
19. Reference 42, compound 1.
20. Reference 42, Figure 11.
21. Reference 56, compound 4.
22. Reference 56, compound 9.
23. Reference 43, compound 5c.
24. Reference 43, compound 1.
25. (2$R$,3$S$)-2,3-dihydroxyhexyl acetate.
26. (2$S$,3$S$)-2,3-dihydroxyhexyl acetate.
27. Reference 60, compound 21.b.
28. Reference 60, compound 18d.
29. Reference 13, compound 33.3.
30. Reference 13, figure 35.
31. Reference 13, compound 40.16.
32. Reference 13, compound 40.1.
33. Reference 13, compound 43.6.
34. Reference 13, compound 43.8.
35. Reference 13, compound 43.8.
36. Reference 13, compound 47.6.
37. Reference 13, compound 47.5.
38. Reference 13, compound 59.3.
39. Reference 13, enantiomer of compound 59.3.
40. Reference 13, compound 59.4.
41. Reference 13, enantiomer of compound 59.1.
42. Reference 13, compound 59.1.
43. Reference 13, compound 59.9.
44. Reference 13, compound 64.4.

45. Reference 13, compound 64.2.
46. Reference 13, enantiomer of compound 64.2.
47. Reference 13, enantiomer of compound 64.7.
48. Reference 13, compound 71.7.
49. Reference 13, compound 71.18.
50. Reference 13, compound 74.1.

# REFERENCES

1. Eliel, E. L.; Wilen, S. H.; Mander, L. N. *Stereochemistry of Organic Compounds*. New York: Wiley-Interscience, 1994.
2. Uray, G. In *Houben-Weyl Methods in Organic Chemistry, Stereoselective Synthesis*, ed. G. Helchen, R. W. Hoffmann, J. Mulzer, E. Schaumann, Workbench Edition E 21, Vol. 1, Chapter 3.2.2, pp. 253–292. Stuttgart, New York: Thieme, 1996.
3. Allenmark, S.; Gawronski, J. "Determination of Absolute Configuration: An Overview." *Chirality* 20 (2008): 606–608.
4. McConnell, O.; Bach, A.; Balibar, C.; Byrne, N.; Cai, Y.; Carter, G.; Chlenov, M.; Di, L.; Fan, K.; Goljer, I.; He, Y.; Herold, D.; Kagan, M.; Kerns, E.; Koehn, F.; Kraml, C.; Marathias, V.; Marquez, B.; McDonald, K.; Nogle, L.; Petucci, C.; Schlingmann, G.; Tawa, G.; Tischler, M.; Williamson, T.; Sutherland, A.; Watts, W.; Young, M.; Zhang, M.; Zhang, Y.; Zhou, D.; Ho, D. "Enantiomeric Separation and Determination of Absolute Stereochemistry of Asymmetric Molecules in Drug Discovery: Building Chiral Technology Toolboxes." *Chirality*, 19 (2007): 658–682.
5. Berova, N; Nakanishi, K; Woody, R. W. *Circular Dichroism: Principles and Applications*, 2nd ed. New York: Wiley-VCH, 2000.
6. Berova, N.; Di Bari, L.; Pescitelli, G. "Application of Electronic Circular Dichroism in Configurational and Conformational Analysis of Organic Compounds." *Chemical Society Reviews* 36 (2007): 914–931.
7. Allenmark, S. G. "Chiroptical Methods in the Stereochemical Analysis of Natural Products (1975 to 1999)." *Natural Product Reports* 17 (2000): 145–155.
8. Flack, H. D.; Bernardinelli, G. "The Use of X-ray Crystallography to Determine Absolute Configuration." *Chirality* 20 (2008): 681–690.
9. Harada, N. "Chiral Auxiliaries Powerful for Both Enantiomer Resolution and Determination of Absolute Configuration by X-ray Crystallography." *Topics in Stereochemistry* 25 (2006): 177–203.
10. Freedman, T. B.; Cao, X.; Dukor, R. K.; Nafie, L. A. "Absolute Configuration Determination of Chiral Molecules in the Solution State Using Vibrational Circular Dichroism." *Chirality* 15 (2003): 743–758.
11. Sherer, E. C.; Lee, C. H.; Shpungin, J.; Cuff, J. F.; Da, C.; Ball, R.; Bach, R.; Crespo, A.; Gong, X. Y.; Welch, C. J. "Systematic Approach to Conformational Sampling for Assigning Absolute Configuration Using Vibrational Circular Dichroism." *Journal of Medicinal Chemistry* 57 (2014): 477–494.
12. Felippe, L. G.; Batista, J. M., Jr.; Baldoqui, D. C.; Nascimento, I. R.; Kato, M. J.; He, Y.; Nafie, L. A.; Furlan, M. "VCD to Determine Absolute Configuration of Natural Product Molecules: Secolignans from Peperomia Blanda." *Organic and Biomolecular Chemistry* 10 (2012): 4208–4214.
13. Seco, J. M.; Quiñoá, E.; Riguera, R. "Assignment of the Absolute Configuration of Polyfunctional Compounds by NMR Using Chiral Derivatizing Agents." *Chemical Reviews* 112 (2012): 4603–4641.

14. Seco, J. M.; Quiñoá, E.; Riguera, R. "The Assignment of Absolute Configuration by NMR." *Chemical Reviews* 140 (2004): 17–117.
15. Seco, J. M.; Quiñoá, E.; Riguera, R. "A Practical Guide for the Assignment of the Absolute Configuration of Alcohols, Amines and Carboxylic Acids by NMR." *Tetrahedron: Asymmetry* 12 (2001): 2915–2925.
16. Dale, J. A.; Mosher, H. S. "Nuclear Magnetic Resonance Nonequivalence of Diastereomeric Esters of α-Substituted Phenylacetic Acids for the Determination of Stereochemical Purity." *Journal of the American Chemical Society* 90 (1968): 3732–3738.
17. Dale, J. A.; Dull, D. L.; Mosher, H. S. "α-Methoxy- α-Trifluoromethylphenylacetic Acid, a Versatile Reagent for the Determination of Enantiomeric Composition of Alcohols and Amines." *Journal of Organic Chemistry* 34 (1969): 2543–2549.
18. Dale, J. A.; Mosher, H. S. "Nuclear Magnetic Resonance Enantiomer Regents. Configurational Correlations Via Nuclear Magnetic Resonance Chemical Shifts of Diastereomeric Mandelate, O-methylmandelate, and α-Methoxy-α-Trifluoromethylphenylacetate (MTPA) Esters." *Journal of the American Chemical Society* 95 (1973): 512–519.
19. Sullivan, G. R.; Dale, J. A.; Mosher, H. S. "Correlation of Configuration and Fluorine-19 Chemical Shifts of α-Methoxy-α-Trifluoromethylphenyl Acetate Derivatives." *Journal of Organic Chemistry* 38 (1973): 2143–2147.
20. Molinski, T. F.; Morinaka, B. I. "Integrated Approaches to the Configurational Assignment of Marine Natural Products." *Tetrahedron* 68 (2012): 9307–9343.
21. Szyszkowiak, J.; Majewska, P. "Determination of Absolute Configuration by 31P NMR." *Tetrahedron: Asymmetry* 25 (2014): 103–112.
22. Blazewska, K. M.; Gajda, T. "Assignment of the Absolute Configuration of Hydroxy- and Aminophosphonates by NMR Spectroscopy." *Tetrahedron: Asymmetry* 20 (2009): 1337–1361.
23. Blazewska, K.; Paneth, P.; Gajda, T. "The Assignment of the Absolute Configuration of Diethyl Hydroxy- and Aminophosphonates by $^1$H and $^{31}$P NMR Using Naproxen as a Reliable Chiral Derivatizing Agent." *Journal of Organic Chemistry* 72 (2007): 878–887.
24. Kwan, E. E.; Huang, Sh. G. "Structural Elucidation with NMR Spectroscopy: Practical Strategies for Organic Chemists." *European Journal of Organic Chemistry* (2008): 2671–2688.
25. Menche, D. "New Methods for Stereochemical Determination of Complex Polyketides: Configurational Assignment of Novel Metabolites from Myxobacteria." *Natural Product Reports* 25 (2008): 905–918.
26. Curran, D. P.; Sui, B. "A 'Shortcut' Mosher Ester Method to Assign Configurations of Stereocenters in Nearly Symmetric Environments: Fluorous Mixture Synthesis and Structure Assignment of Petrocortyne A." *Journal of the American Chemical Society* 131 (2009): 5411–5413.
27. Kusumi, T.; Ooi, T.; Ohkubo, Y.; Yabuuchi, T. "The Modified Mosher's Method and the Sulfoximine Method." *Bulletin of the Chemical Society of Japan* 79 (2006): 965–980.
28. Higashibayashi, S.; Czechtizky, W.; Kobayashi, Y.; Kishi, Y. "Universal NMR Databases for Contiguous Polyols." *Journal of the American Chemical Society* 125 (2003): 14379–14393.

29. Hoye, T. R.; Erickson, S. E.; Erickson-Birkedahl, Sh. L.; Hale, C. R. H.; Izgu, E. C.; Mayer, M. J.; Notz, P. K.; Renner, M. K. "Long-Range Shielding Effects in the $^1$H NMR Spectra of Mosher-Like Ester Derivatives." *Organic Letters* 12 (2010): 1768–1771.
30. Zhang, QY; Carrera, G.; Gomes, M. J. S.; Aires-de-Sousa, J. "Automatic Assignment of Absolute Configuration from 1D NMR Data." *Journal of Organic Chemistry* 70 (2005): 2120–2130.
31. Berger, R.; Courtieu, J.; Gil, R. R.; Griesinger, C.; Köck, M.; Lesot, P.; Luy, B.; Merlet, D.; Navarro-Vázquez, A.; Reggelin, M.; Reinscheid, U. M.; Thiele, C. M.; Zweckstetter, M. "Is Enantiomer Assignment Possible by NMR Spectroscopy Using Residual Dipolar Couplings from Chiral Nonracemic Alignment Media? A Critical Assessment." *Angewandte Chemie, International Edition* 51 (2012): 8388–8391.
32. Kummerloewe, G.; Luy, B. "Residual Dipolar Couplings for the Configurational and Conformational Analysis of Organic Molecules." *Annual Reports on NMR Spectroscopy* 68 (2009): 193–232.
33. Wenzel, T. J. *Discrimination of Chiral Compounds Using NMR Spectroscopy*. Hoboken, NJ: Wiley, 2007.
34. Wenzel, T. J.; Chisholm, C. D. "Assignment of Absolute Configuration Using Chiral Reagents and NMR Spectroscopy." *Chirality* 23 (2011): 190–214.
35. Seco, J. M.; Latypov, Sh.; Quiñoá, E.; Riguera, R. "New Chirality Recognizing Reagents for the Determination of Absolute Stereochemistry and Enantiomeric Purity by NMR." *Tetrahedron Letters* 35 (1994): 2921–2924.
36. Seco, J. M.; Latypov, Sh.; Quiñoá, E.; Riguera, R. "Determination of the Absolute Configuration of Alcohols by Low Temperature $^1$H NMR of Aryl(Methoxy)Acetates." *Tetrahedron: Asymmetry* 61 (1995): 107–110.
37. Latypov, Sh. K.; Seco, J. M.; Quiñoá, E.; Riguera, R. "Conformational Structure and Dynamics of Arylmethoxyacetates: DNMR Spectroscopy and Aromatic Shielding Effect" *Journal of Organic Chemistry* 60 (1995): 504–515.
38. Latypov, Sh. K.; Seco, J. M.; Quiñoá, E.; Riguera, R. "MTPA vs. MPA in the Determination of the Absolute Configuration of Chiral Alcohols by $^1$H NMR." *Journal of Organic Chemistry* 61(1996): 8569–8577.
39. Seco, J. M.; Latypov, Sh. K.; Quiñoá, Emilio; Riguera, R. "Determining Factors in the Assignment of the Absolute Configuration of Alcohols by NMR: The Use of Anisotropic Effects on Remote Positions." *Tetrahedron* 53 (1997): 8541–8564.
40. Seco, J. M.; Tseng, L.; Godejohann, M.; Quiñoá, E.; Riguera, R. "Simultaneous Enantioresolution and Assignment of Absolute Configuration of Secondary Alcohols by Directly Coupled HPLC-NMR of 9-AMA Esters." *Tetrahedron: Asymmetry* 13 (2002): 2149–2153.
41. Latypov, Sh. K.; Seco, J. M.; Quiñoá, E.; Riguera, R. "Are Both the (R)- and the (S)-MPA Esters Really Needed for the Assignment of the Absolute Configuration of Secondary Alcohols by NMR? The Use of a Single Derivative." *Journal of the American Chemical Society* 120 (1998): 877–882.
42. García, R.; Seco, J. M.; Vázquez, S. A.; Quiñoá, E.; Riguera, R. "Absolute Configuration of Secondary Alcohols by $^1$H NMR: In Situ Complexation of α-Methoxyphenylacetic Acid Esters With Barium(II)." *Journal of Organic Chemistry* 67 (2002): 4579–4589.

43. Seco, J. M.; Quiñoá, E.; Riguera, R. "9-Anthrylmethoxyacetic Acid Esterification Shifts-correlation with the Absolute Stereochemistry of Secondary Alcohols." *Tetrahedron* 55 (1999): 569–584.
44. Ferreiro, M. J.; Latypov, Sh. K.; Quiñoá, E.; Riguera, R. "Determination of the Absolute Configuration and Enantiomeric Purity of Chiral Primary Alcohols by $^1$H NMR of 9-Anthrylmethoxyacetates." *Tetrahedron: Asymmetry* 7 (1996): 2195–2198.
45. Latypov, Sh. K.; Ferreiro, M. J.; Quiñoá, E.; Riguera, R. "Assignment of the Absolute Configuration of β-Chiral Primary Alcohols by NMR: Scope and Limitations." *Journal of the American Chemical Society* 120 (1998): 4741–4751.
46. Freire, F.; Seco, J. M.; Quiñoá, E.; Riguera, R. "Challenging the Absence of Observable Hydrogens in the Assignment of Absolute Configurations by NMR: Application to Chiral Primary Alcohols." *Chemical Communications* (2007): 1456–1458.
47. Louzao, I.; Seco, J. M.; Quiñoá, E.; Riguera, R. "The Assignment of Absolute Configuration of Cyanohydrins by NMR." *Chemical Communications* (2006): 1422–1424.
48. Louzao, I.; García, R.; Seco, J. M.; Quiñoá, E.; Riguera, R. "Absolute Configuration of Ketone Cyanohydrins by $^1$H NMR: The Special Case of Polar Substituted Tertiary Alcohols." *Organic Letters* 11 (2009): 53–56.
49. Louzao, I.; Seco, J. M.; Quiñoá, E.; Riguera, R. "The Use of a Single Derivative in the Configurational Assignment of Ketone Cyanohydrins." *European Journal of Organic Chemistry* (2010): 6520–6524.
50. Porto, S.; Seco, J. M.; Ortiz, A.; Quiñoá, E.; Riguera, R. "Chiral Thiols: The Assignment of their Absolute Configuration by $^1$H NMR." *Organic Letters* 9 (2007): 5015–5018.
51. Porto, S.; Quiñoá, E.; Riguera, R. "Designing Chiral Derivatizing Agents (CDA) for the NMR Assignment of the Absolute Configuration: A Theoretical and Experimental Approach with Thiols as a Case Study." *Tetrahedron* 70 (2014): 3276–3283.
52. Latypov, Shamil K.; Seco, J. M.; Quiñoá, E.; Riguera, R. "Determination of the Absolute Stereochemistry of Chiral Amines by $^1$H NMR of Arylmethoxyacetic Acid Amides: The Conformational Model." *Journal of Organic Chemistry* 60 (1995): 1538–1545.
53. Seco, J. M.; Latypov, Sh. K.; Quiñoá, E.; Riguera, R. "Choosing the Right Reagent for the Determination of the Absolute Configuration of Amines by NMR: MTPA or MPA?" *Journal of Organic Chemistry* 62 (1997): 7569–7574.
54. Seco, J. M.; Quiñoá, E.; Riguera, R. "Boc-Phenylglycine: The Reagent of Choice for the Assignment of the Absolute Configuration of α-Chiral Primary Amines by $^1$H NMR Spectroscopy." *Journal of Organic Chemistry* 64 (1999): 4669–4675.
55. Lopez, B.; Quiñoá, E.; Riguera, R. "Complexation with Barium(II) Allows the Inference of the Absolute Configuration of Primary Amines by NMR." *Journal of the American Chemical Society* 121 (1999): 9724–9725.
56. García, R.; Seco, J. M.; Vázquez, S. A.; Quiñoá, E.; Riguera, R. "Role of Barium(II) in the Determination of the Absolute Configuration of Chiral Amines by $^1$H NMR Spectroscopy." *Journal of Organic Chemistry* 71 (2006): 1119–1130.
57. Ferreiro, M. J.; Latypov, Sh. K.; Quiñoá, E.; Riguera, R. "Assignment of the Absolute Configuration of α-Chiral Carboxylic Acids by $^1$H NMR Spectroscopy." *Journal of Organic Chemistry* 65 (2000): 2658–2666.
58. Ferreiro, M. J.; Latypov, Sh. K.; Quiñoá, E.; Riguera, R. "The Use of Ethyl 2-(9-Anthryl)-2-Hydroxyacetate for Assignment of the Absolute Configuration of Carboxylic Acids by $^1$H NMR." *Tetrahedron: Asymmetry* 8 (1997): 1015–1018.

59. Seco, J. M.; Martino, M.; Quiñoá, E.; Riguera, R. "Absolute Configuration of 1,n-Diols by NMR: The Importance of the Combined Anisotropic Effects in Bis-Arylmethoxyacetates." *Organic Letters* 2 (2000): 3261–3264.
60. Freire, F.; Seco, J. M.; Quiñoá, E.; Riguera, R. "Determining the Absolute Stereochemistry of Secondary/Secondary Diols by $^1$H NMR: Basis and Applications." *Journal of Organic Chemistry* 70 (2005): 3778–3790.
61. Freire, F.; Calderon, F.; Seco, J. M.; Fernandez-Mayoralas, A.; Quiñoá, E.; Riguera, R. "Relative and Absolute Stereochemistry of Secondary/Secondary Diols: Low-Temperature $^1$H NMR of their Bis-MPA Esters." *Journal of Organic Chemistry* 72 (2007): 2297–2301.
62. Leiro, V.; Freire, F.; Quiñoá, E.; Riguera, R. "Absolute Configuration of Amino Alcohols by $^1$H-NMR." *Chemical Communications* (2005): 5554–5556.
63. Freire, F.; Seco, J. M.; Quiñoá, E.; Riguera, R. "The Prediction of the Absolute Stereochemistry of Primary and Secondary 1,2-Diols by $^1$H NMR Spectroscopy: Principles and Applications." *Chemistry: A European Journal* 11 (2005): 5509–5522.
64. Freire, F.; Seco, J. M.; Quiñoá, E.; Riguera, R. "Chiral 1,2-Diols: The Assignment of Their Absolute Configuration by NMR Made Easy." *Organic Letters* 12 (2010): 208–211.
65. Freire, F.; Seco, J. M.; Quiñoá, E.; Riguera, R. "The Assignment of the Absolute Configuration of 1,2-Diols by Low-Temperature NMR of a Single MPA Derivative." *Organic Letters* 7 (2005): 4855–4858.
66. Leiro, V.; Seco, J. M.; Quiñoá, E.; Riguera, R. "Cross Interaction Between Auxiliaries: The Chirality of Amino Alcohols by NMR." *Organic Letters* 10 (2008): 2729–2732.
67. Leiro, V.; Seco, J. M.; Quiñoá, E.; Riguera, R. "Using a Combination of Magnetic Anisotropic Effects for the Configurational Assignment of Amino Alcohols. " *Chemistry: An Asian Journal* 5 (2010): 2106–2112.
68. Leiro, V.; Seco, J. M.; Quiñoá, E.; Riguera, R. "Assigning the Configuration of Amino Alcohols by NMR: A Single Derivatization Method." *Organic Letters* 10 (2008): 2733–2736.
69. Lallana, E.; Freire, F.; Seco, J. M.; Quiñoá, E.; Riguera, R. "The $^1$H NMR Method for the Determination of the Absolute Configuration of 1,2,3-Prim,Sec,Sec-Triols." *Organic Letters* 8 (2006): 4449–4452.
70. Freire, F.; Lallana, E.; Quiñoá, E.; Riguera, R. "The Stereochemistry of 1,2,3-Triols Revealed by $^1$H NMR Spectroscopy: Principles and Applications." *Chemistry: A European Journal* 15 (2009): 11963–11975.
71. Frisch, M. J.; Trucks, G. W.; Schlegel, H. B.; Scuseria, G. E.; Robb, M. A.; Cheeseman, J. R.; Montgomery, Jr., J. A.; Vreven, T.; Kudin, K. N.; Burant, J. C.; Millam, J. M.; Iyengar, S. S.; Tomasi, J.; Barone, V.; Mennucci, B.; Cossi, M.; Scalmani, G.; Rega, N.; Petersson, G. A.; Nakatsuji, H.; Hada, M.; Ehara, M.; Toyota, K.; Fukuda, R.; Hasegawa, J.; Ishida, M.; Nakajima, T.; Honda, Y.; Kitao, O.; Nakai, H.; Klene, M.; Li, X.; Knox, J. E.; Hratchian, H. P.; Cross, J. B.; Bakken, V.; Adamo, C.; Jaramillo, J.; Gomperts, R.; Stratmann, R. E.; Yazyev, O.; Austin, A. J.; Cammi, R.; Pomelli, C.; Ochterski, J. W.; Ayala, P. Y.; Morokuma, K.; Voth, G. A.; Salvador, P.; Dannenberg, J. J.; Zakrzewski, V. G.; Dapprich, S.; Daniels, A. D.; Strain, M. C.; Farkas, O.; Malick, D. K.; Rabuck, A. D.; Raghavachari, K.; Foresman, J. B.; Ortiz, J. V.; Cui, Q.; Baboul, A. G.; Clifford, S.; Cioslowski, J.; Stefanov, B. B.; Liu, G.; Liashenko, A.; Piskorz, P.; Komaromi, I.; Martin, R. L.; Fox, D. J.; Keith, T.; Al-Laham, M. A.; Peng, C. Y.;

Nanayakkara, A.; Challacombe, M.; Gill, P. M. W.; Johnson, B.; Chen, W.; Wong, M. W.; Gonzalez, C.; and Pople, J. A. Gaussian 03, Revision E.01; Gaussian, Inc., Wallingford, CT, 2004.

72. Louzao, I.; Seco, J. M.; Quiñoá, E.; Riguera, R. "[13]C NMR as a General Tool for the Assignment of Absolute Configuration." *Chemical Communications* 46 (2010): 7903–7905.

73. Kobayashi, M. "The [13]C NMR Method for Determining the Absolute Configuration of the 1,2-Glycols Consisting of Secondary and Tertiary Hydroxyl Groups." *Tetrahedron* 56 (2000): 1661–1665.

74. Arnone, A.; Bernardi, R.; Blasco, F.; Cardillo, R.; Resnati, G.; Gerus, I. I.; Kukhar, V. P. "Trifluoromethyl vs. Methyl Ability to Direct Enantioselection in Microbial Reduction of Carbonyl Substrates." *Tetrahedron* 54 (1998): 2809–2818.

75. Fukushi, Y.; Shigematsu, K.; Mizutani, J.; Tahara, S. "A New NMR Chiral Derivatizing Reagent for Determining the Absolute Configurations of Carboxylic Acids." *Tetrahedron Letters* 37 (1996): 4737–4740.

76. Fukushi, Y.; Yajima, C.; Mizutani, J. "A New Method for Establishment of Absolute Configurations of Secondary Alcohols by NMR Spectroscopy." *Tetrahedron Letters* 35 (1994): 599–602.

77. Pehk, T.; Lippmaa, E.; Lopp, M.; Paju, A.; Borer, B. C.; Taylor, R. J. K. "Determination of the Absolute Configuration of Chiral Secondary Alcohols: New Advances Using Carbon 13 and 2D-NMR Spectroscopy." *Tetrahedron: Asymmetry* 4 (1993): 1527–1532.

78. Muñoz, M. A.; Joseph-Nathan, P. "DFT-GIAO [1]H and [13]C NMR Prediction of Chemical Shifts for the Configurational Assignment of 6β-Hydroxyhyoscyamine Diastereoisomers." *Magnetic Resonance in Chemistry* 47 (2009): 578–584.

79. Wiitala, K. W.; Cramer, C. J.; Hoye, T. R. "Comparison of Various Density Functional Methods For Distinguishing Stereoisomers Based on Computed [1]H or [13]C NMR Chemical Shifts Using Diastereomeric Penam β-Lactams as a Test Set." *Magnetic Resonance in Chemistry* 45 (2007): 819–829.

80. Pérez-Trujillo, M.; Monteagudo, E.; Parella, T. "[13]C NMR Spectroscopy for the Differentiation of Enantiomers Using Chiral Solvating Agents." *Analytical Chemistry* 85 (2013): 10887–10894.

81. Kalinovsky, H.-O.; Berger, S.; Braun, S. *Carbon-13 NMR Spectroscopy.* Salisbury: John Wiley and Sons, 1986, p. 92.

82. Seco, J. M.; Quiñoá, E.; Riguera, R. "The Assignment of Absolute Configurations by NMR of Arylmethoxyacetate Derivatives: Is this Methodology Being Correctly Used?" *Tetrahedron: Asymmetry* 11 (2000): 2781–2791.

83. Porto, S.; Duran, J.; Seco, J. M.; Quiñoá, E.; Riguera, R. "Mix and Shake Method for Configurational Assignment by NMR: Application to Chiral Amines and Alcohols." *Organic Letters* 5 (2003): 2979–2982.

84. Porto, S.; Seco, J. M.; Espinosa, J. F.; Quiñoá, E.; Riguera, R. "Resin-Bound Chiral Derivatizing Agents for Assignment of Configuration by NMR Spectroscopy." *Journal of Organic Chemistry* 73 (2008): 5714–5722.

85. Joshi, B. S.; Pelletier, S. W. "A Cautionary Note on the Use of Commercial (R)-MTPA-Cl and (S)-MTPA-Cl in Determination of Absolute Configuration by Mosher Ester Analysis." *Heterocycles* 51(1999): 183–184.

86. Pham, V. C.; Jossang, A.; Sevenet, T.; Nguyen, V. H.; Bodo, B. "Myrioneurinol: A Novel Alkaloid Skeleton from Myrioneuron Nutans." *Tetrahedron* 63 (2007): 11244–11249.

87. Moreno-Dorado,F.J.;Guerra,F.M.;Ortega,M.J.;Zubía,E.;Massanet,G.M. "Enantioselective Synthesis of Arylmethoxyacetic Acid Derivatives." *Tetrahedron: Asymmetry* 14 (2003): 503–510.
88. Li, B. L.; Zhang, Z. G.; Du, L. L.; Wang, W. "Chiral Resolutions of (9-Anthryl) Methoxyacetic Acid and (9-Anthryl)Hydroxyacetic Acid by Capillary Electrophoresis." *Chirality* 39 (2008): 2035–2039.
89. Arita, S.; Yabuuchi, T.; Kusumi, T. "Resolution of 1- and 2-Naphthylmethoxyacetic Acids, NMR Reagents for Absolute Configuration Determination, by Use of L-Phenylalaninol." *Chirality* 15 (2003): 609–614.
90. Kimura, M.; Kuboki, A.; Sugai, T. "Chemo-Enzymatic Synthesis of Enantiomerically Pure (R)-2-Naphthylmethoxyacetic Acid." *Tetrahedron: Asymmetry* 13 (2002): 1059–1068.
91. Trost, B. M.; Belletire, J. L.; Godleski, S.; McDougal, P. G.; Balkovec, J. M. "On the Use of the O-Methylmandelate Ester for Establishment of Absolute Configuration of Secondary Alcohols." *Journal of Organic Chemistry* 51 (1986): 2370–2374.
92. Abad, J. L.; Fabrias, G.; Camps, F. "Synthesis of Dideuterated and Enantiomers of Monodeuterated Tridecanoic Acids at C-9 and C-10 Positions." *Journal of Organic Chemistry* 65 (2000): 8582–8588.
93. Ward, D. E.; Rhee, C. K. "A Simple Method for the Miscroscale Preparation of Mosher Acid Chloride." *Tetrahedron Letters* 32 (1991): 7165–7166.
94. Smith, R. A.; Bhargava, A.; Browe, C.; Chen, J. S.; Dumas, J.; Hatoum-Mokdad, H.; Romero, R. "Discovery and Parallel Synthesis of a New Class of Cathepsin K Inhibitors." *Bioorganic and Medicinal Chemistry Letters* 11 (2001): 2951–2954.
95. Nicolaou, K. C.; Pfefferkon, J. A.; Barluenga, S.; Mitchel, H. J.; Roecker, A. J.; Cao, G. Q. " Natural Product-Like Combinatorial Libraries Based on Privileged Structures. 3: The "Libraries from Libraries" Principle for Diversity Enhancement of Benzopyran Libraries." *Journal of the American Chemical Society* 122 (2000): 9968–9976.
96. Takayanagi, M.; Flessner, T.; Wong, C. H. "A Strategy for the Solution-Phase Parallel Synthesis of N-(Pyrrolidinylmethyl)Hydroxamic Acids." *Journal of Organic Chemistry* 65 (2000): 3811–3815.
97. Tajima, H.; Wakimoto, T.; Takada, K.; Ise, Y.; Abe, I. "Revised Structure of Cyclolithistide A, a Cyclic Depsipeptide from the Marine Sponge Discodermia Japonica." *Journal of Natural Products* 77 (2014): 154–158.
98. Irschik, H.; Washausen, P.; Sasse, F.; Fohrer, J.; Huch, V.; Mueller, R.; Prusov, E. V. "Isolation, Structure Elucidation, and Biological Activity of Maltepolides: Remarkable Macrolides from Myxobacteria." *Angewandte Chemie, International Edition* 52 (2013): 5402–5405.
99. Xiaochao, X.; Zhaojun, Y.; Xiangbao, M.; Zhongjun, L. "A Carbohydrate-Based Approach for the Total Synthesis of (−)-Dinemasone B, (+)-4a-Epi-Dinemasone B, (−)-7-Epi-Dinemasone B, and (+)-4a,7-Di-Epi-Dinemasone B." *Journal of Organic Chemistry* 78 (2013): 9354–9365.
100. Sun, P.; Xu, D. X.; Mandi, A.; Kurtan, T.; Li, T. J.; Schulz, B.; Zhang, W. "Structure, Absolute Configuration, and Conformational Study of 12-Membered Macrolides from the Fungus Dendrodochium Sp. Associated with the Sea Cucumber Holothuria Nobilis Selenka." *Journal of Organic Chemistry* 78 (2013): 7030–7047.
101. Ren, J.; Liu, D.; Tian, L; Wei, Y.; Proksch, P.; Zeng, J.; Lin, W. "Venezuelines A-G, New Phenoxazine-Based Alkaloids and Aminophenols from Streptomyces Venezuelae and

the Regulation of Gene Target Nur77." *Bioorganic and Medicinal Chemistry Letters* 23 (2013): 301–304.
102. Chen, M.; Lin, S.; Li, L.; Zhu, C.; Wang, X.; Wang, Y.; Jiang, B.; Wang, S.; Li, Y.; Jiang, J.; Shi, J. "Enantiomers of an Indole Alkaloid Containing Unusual Dihydrothiopyran and 1,2,4-Thiadiazole Rings from the Root of Isatis Indigotica." *Organic Letters* 14 (2012): 5668–5671.
103. Tello, E.; Castellanos, L.; Arevalo-Ferro, C.; Rodriguez, J.; Jimenez, C.; Duque, C. "Absolute Stereochemistry of Antifouling Cembranoid Epimers at C-8 from the Caribbean Octocoral Pseudoplexaura Flagellosa: Revised Structures of Plexaurolones." *Tetrahedron* 67 (2011): 9112-9121.
104. Mayorga, H.; Urrego, N. F.; Castellanos, L.; Duque, C. "Cembradienes from the Caribbean Sea Whip Eunicea Sp." *Tetrahedron Letters* 52 (2011): 2515–2518.
105. Lai, D.; Li, Yongxin; Minjuan, X.; Deng, Z.; Van Ofwegen, L.; Qian, P.; Proksch, P.; Lin, W. "Sinulariols A-S, 19-Oxygenated Cembranoids from the Chinese Soft Coral Sinularia Rigida." *Tetrahedron* 67 (2011): 6018–6029.
106. Igarashi, Y.; Kim, Y.; In, Y.; Ishida, T.; Kan, Y.; Fujita, T.; Iwashita, T.; Tabata, H.; Onaka, H.; Furumai, T. "Alchivemycin A, a Bioactive Polycyclic Polyketide with an Unprecedented Skeleton from Streptomyces Sp." *Organic Letters* 12 (2010): 3402–3405.
107. De Berardinis, A M.; Turlington, M; Ko, J; Sole, L; Pu, L. "Facile Synthesis of a Family of H8BINOL-Amine Compounds and Catalytic Asymmetric Arylzinc Addition to Aldehydes." *Journal of Organic Chemistry* 74 (2010): 2836–2850.
108. Alcaide, B.; Almendros, P.; Carrascosa, R.; Martínez del Campo, T. "Metal-Catalyzed Cycloetherification Reactions of β,γ- and γ,δ-Allendiols: Chemo-, Regio-, and Stereocontrol in the Synthesis of Oxacycles." *Chemistry: A European Journal* 16 (2010): 13243–13252.
109. Ordóñez, M.; Labastida-Galvan, V.; Lagunas-Rivera, S. "Stereoselective Synthesis of GABOB, Carnitine and Statine Phosphonates Analogues." *Tetrahedron: Asymmetry* 21 (2010): 129–147.
110. Millan-Aguinaga, N.; Soria-Mercado, I. E.; Williams, P. "Xestosaprol D and E from the Indonesian Marine Sponge Xestospongia Sp." *Tetrahedron Letters* 51 (2010): 751–753.
111. Tello, E.; Castellanos, L.; Arevalo-Ferro, C.; Duque, C. "Cembranoid Diterpenes from the Caribbean Sea Whip Eunicea Knighti." *Journal of Natural Products* 72 (2009): 1595–1602.
112. Yin, Y. Q.; Wang, J. S.; Luo, J. G.; Kong, L. Y. "Novel Acylated Lipo-Oligosaccharides from the Tubers of Ipomoea Batatas." *Carbohydrate Research* (2009): 466–473.
113. Pontius, A.; Krick, A.; Kehraus, S.; Brun, R.; Koenig, G. M. "Antiprotozoal Activities of Heterocyclic-Substituted Xanthones from the Marine-Derived Fungus Chaetomium Sp." *Journal of Natural Products* 71 (2008): 1579–1584.
114. Zubía, E.; Ortega, M. J.; Hernández-Guerrero, C. J.; Carballo, J. L. "Isothiocyanate Sesquiterpenes from a Sponge of the Genus Axinyssa." *Journal of Natural Products* (2008): 608–614.
115. Pontius, A.; Krick, A.; Kehraus, S.; Foegen, S. E.; Mueller, M.; Klimo, K.; Gerhaeuser, C.; Koenig, G. M. "Noduliprevenone: A Novel Heterodimeric Chromanone With Cancer Chemopreventive Potential." *Chemistry: A European Journal* 14 (2008): 9860–9863.
116. Maerten, E.; Agbossou-Niedercorn, F.; Castanet, Y.; Mortreux, A. "Preparation of Pyridinyl Aryl Methanol Derivatives by Enantioselective Hydrogenation of Ketones

Using Chiral Ru(diphosphine)(diamine) Complexes: Attribution of their Absolute Configuration by ¹H NMR Spectroscopy Using Mosher's Reagent." *Tetrahedron* 64 (2008): 8700–8708.

117. Zubía, E.; Ortega, M. J.; Carballo, J. L. "Sesquiterpenes from the Sponge Axinyssa Isabela." *Journal of Natural Products* 71 (2008): 2004–2010.

118. Lemos, E., de; Poree, F. H.; Bourin, A.; Barbion, J.; Agouridas, E.; Lannou, M. I.; Commercon, Alain; Betzer, J. F.; Pancrazi, A.; Ardisson, J. "Total Synthesis of Discodermolide: Optimization of the Effective Synthetic Route." *Chemistry: A European Journal* 14 (2008): 11092–11112.

119. Suárez-Ortiz, G. A.; Cerda-García-Rojas, C. M.; Hernández-Rojas, A.; Pereda-Miranda, R. "Absolute Configuration and Conformational Analysis of Brevipolides, Bioactive 5,6-Dihydro-α-Pyrones from Hyptis Brevipes." *Journal of Natural Products* 76 (2013): 72–78.

120. Nam, S. J.; Kauffman, C. A.; Paul, L. A.; Jensen, P. R.; Fenical, W. "Actinoranone, a Cytotoxic Meroterpenoid of Unprecedented Structure from a Marine Adapted Streptomyces Sp." *Organic Letters* 15 (2013): 5400–5403.

121. Carrera, I.; Brovetto, M.; Seoane, G. A. "Chemoenzymatic Preparation of (6R)-5,6-Dihydro-2H-Pyran-2-One: A Ubiquitous Structural Motif of Biologically Active Lactones." *Tetrahedron: Asymmetry* 24 (2013): 1467–1472.

122. Altendorfer, M.; Raja, A.; Sasse, F.; Irschik, H.; Menche, D. "Modular Synthesis of Polyene Side Chain Analogues of the Potent Macrolide Antibiotic Etnangien by a Flexible Coupling Strategy Based on Hetero-Bis-Metallated Alkenes." *Organic and Biomolecular Chemistry* 11 (2013): 2116–2139.

123. Guzii, A. G.; Makarieva, T. N.; Korolkova, Y. V.; Andreev, Ya. A.; Mosharova, I. V.; Tabakmaher, K. M.; Denisenko, V. A.; Dmitrenok, P. S.; Ogurtsova, E. K.; Antonov, A. S.; Leec, H. S.; Grishinb, E. V. "Pulchranin A, Isolated from the Far-Eastern Marine Sponge, Monanchora Pulchra: The First Marine Non-Peptide Inhibitor of TRPV-1 Channels." *Tetrahedron Letters* 54 (2013): 1247–1250.

124. Das, S.; Goswami, R. K. "Stereoselective Total Synthesis of Ieodomycins A and B and Revision of the NMR Spectroscopic Data of Ieodomycin B." *Journal of Organic Chemistry* 78 (2013): 7274–7280.

125. Ioannou, E.; Quesada, A.; Rahman, M. M.; Gibbons, S.; Vagias, C.; Roussis, V. "Structures and Antibacterial Activities of Minor Dolabellanes from the Brown Alga Dilophus Spiralis." *European Journal of Organic Chemistry* 27 (2012): 5177–5186.

126. Dao, T. T.; Tran, T. L.; Kim, J.; Nguyen, P. H.; Lee, E. H.; Park, J.; Jang, I. S.; Oh, W. K. "Terpenylated Coumarins as SIRT1 Activators Isolated from Ailanthus Altissima." *Journal of Natural Products* 75 (2012): 1332–1338.

127. Choi, H.; Proteau, P. J.; Byrum, T.; Gerwick, W. H. "Cymatherelactone and Cymatherols A–C, Polycyclic Oxylipins from the Marine Brown Alga Cymathere Triplicata Original Research Article." *Phytochemistry* 73 (2012): 134–141.

128. Hickmann, V.; Kondoh, A.; Gabor, B.; Alcarazo, M.; Furstner, A. "Catalysis-Based and Protecting-Group-Free Total Syntheses of the Marine Oxylipins Hybridalactone and the Ecklonialactones A, B, and C." *Journal of the American Chemical Society* 133 (2011): 13471–13480.

129. Wierzejska, J.; Ohshima, M.; Inuzuka, T.; Sengoku, T.; Takahashi, M.; Yoda, H. "Total Synthesis and Absolute Stereochemistry of (+)-Batzelasside B and its C8-Epimer, a

New Class of Piperidine Alkaloids from the Sponge Batzella Sp." *Tetrahedron Letters* 52 (2011): 1173–1175.

130. Chen, Y. H.; Tai, C. Y.; Hwang, T. L.; Weng, C. F.; Li, J. J.; Fang, L. S.; Wang, W. H.; Wu, Y. C.; Sung, P. J. "Cladielloides A and B: New Eunicellin-Type Diterpenoids from an Indonesian Octocoral Cladiella Sp." *Marine Drugs* 8 (2010): 2936–2945.

131. Lhullier, C.; Falkenberg, M.; Ioannou, E.; Quesada, A.; Papazafiri, P.; Horta, P. A.; Schenkel, E. P.; Vagias, C.; Roussis, V. "Cytotoxic Halogenated Metabolites from the Brazilian Red Alga Laurencia Catarinensis." *Journal of Natural Products* 73 (2010): 27–32.

132. Smyrniotopoulos, V.; Vagias, C.; Bruyere, C.; Lamoral-Theys, D.; Kiss, R.; Roussis, V. "Structure and In Vitro Antitumor Activity Evaluation of Brominated Diterpenes from the Red Alga Sphaerococcus Coronopifolius." *Bioorganic and Medicinal Chemistry* 18 (2010): 1321–1330.

133. Denmark, S. E.; Kobayashi, T.; Regens, C. S. "Total Synthesis of (+)-Papulacandin D." *Tetrahedron* 66 (2010): 4745–4759.

134. Wacharasindhu, S.; Worawalai, W.; Rungprom, W.; Phuwapraisirisan, P. "(+)-Proto-Quercitol, a Natural Versatile Chiral Building Block for the Synthesis of the α-Glucosidase Inhibitors, 5-Amino-1,2,3,4-Cyclohexanetetrols." *Tetrahedron Letters* 50 (2009): 2189–2192.

135. Kim, M. Y.; Sohn, J. H.; Ahn, J. S.; Oh, H. "Alternaramide, a Cyclic Depsipeptide from the Marine-Derived Fungus Alternaria Sp. SF-5016." *Journal of Natural Products* 72 (2009): 2065–2068.

136. Bindl, M.; Jean, L.; Herrmann, J.; Mueller, R.; Fuerstner, A. "Preparation, Modification, and Evaluation of Cruentaren A and Analogues." *Chemistry: A European Journal* 15 (2009): 12310–12319.

137. Rohr, K.; Herre, R.; Mahrwald, R. "Toward Asymmetric Aldol-Tishchenko Reactions with Enolizable Aldehydes: Access to Defined Configured Stereotriads, Tetrads, and Stereopentads." *Journal of Organic Chemistry* 74 (2009): 3744–3749.

138. Ioannou, E.; Quesada, A.; Vagias, C.; Roussis, V. "Dolastanes from the Brown Alga Dilophus Spiralis: Absolute Stereochemistry and Evaluation of Cytotoxicity." *Tetrahedron* 64 (2008): 3975–3979.

139. Ioannou, E.; Abdel-Razik, A. F.; Alexi, X.; Vagias, C.; Alexis, Michael N.; Roussis, V. "Pregnanes with Antiproliferative Activity from the Gorgonian Eunicella Cavolini Tetrahedron." *Tetrahedron* 64 (2008): 11797–11801.

140. Bock, M.; Buntin, K.; Mueller, R.; Kirschning, A. "Stereochemical Determination of Thuggacins A-C, Highly Active Antibiotics from the Mycobacterium Sorangium Cellulosum." *Angewandte Chemie, International Edition* 47 (2008): 2308–2311.

141. Ahmed, S. A.; Ross, S. A.; Slade, D.; Radwan, M. M.; Khan, I. A.; ElSohly, M. A. "Structure Determination and Absolute Configuration of Cannabichromanone Derivatives from High Potency Cannabis Sativa." *Tetrahedron Letters* 49 (2008): 6050–6053.

142. Webb, M. R.; Addie, M. S.; Crawforth, C. M.; Dale, J. W.; Franci, X.; Pizzonero, M.; Donald, C.; Taylor, R. J. K. "The Syntheses of Rac-Inthomycin A, (+)-Inthomycin B and (+)-Inthomycin C Using a Unified Synthetic Approach." *Tetrahedron* 64 (2008): 4778–4791.

143. Gutierrez, M.; Pereira, A. R.; Debonsi, H. M.; Ligresti, A.; Di Marzo, V.; Gerwick, W. H. "Cannabinomimetic Lipid from a Marine Cyanobacterium." *Journal of Natural Products* 74 (2011): 2313–2317.

144. Probst, N. P.; Haudrechy, A.; Ple, K. "Directed Diastereoselectivity in the Asymmetric Claisen/Metathesis Reaction Sequence." *Journal of Organic Chemistry* 73 (2008): 4338–4341.
145. Ciminiello, P.; Dell'Aversano, C.; Fattorusso, E.; Forino, M.; Magno, S.; Santelia, F. U.; Moutsos, V. I.; Pitsinos, E. N.; Couladouros, E. A. "Oxazinins from Toxic Mussels: Isolation of a Novel Oxazinin and Reassignment of the C-2 Configuration of Oxazinin-1 and -2 on the Basis of Synthetic Models." *Tetrahedron* 62 (2006): 7738–7743.
146. Ciminiello, P.; Dell'Aversano, C.; Fattorusso, C.; Fattorusso, E.; Forino, M.; Magno, S. "Assignment of the Absolute Stereochemistry of Oxazinin-1: Application of the 9-AMA Shift-correlation Method for β-chiral Primary Alcohols." *Tetrahedron* 57 (2001): 8189–8192.
147. Alcaide, B.; Almendros, P.; Cabrero, G.; Ruiz, M. P. "Stereocontrolled Access to Orthogonally Protected Anti,Anti-4-Aminopiperidine-3,5-Diols Through Chemoselective Reduction of Enantiopure β-Lactam Cyanohydrins." *Journal of Organic Chemistry* 72 (2007): 7980–7991.
148. Tasker, S. Z.; Bosscher, M. A.; Shandro, C. A.; Lanni, E. L.; Ryu, K. A.; Snapper, G. S.; Utter, J. M.; Ellsworth, B. A.; Anderson, C. E. "Preparation of N-Alkyl 2-Pyridones via a Lithium Iodide Promoted O- to N-Alkyl Migration: Scope and Mechanism." *Journal of Organic Chemistry* 77 (2012): 8220–8230.
149. Ito, Y.; Ishida, K.; Okada, S.; Murakami, M. "The Absolute Stereochemistry of Anachelins, Siderophores from the Cyanobacterium Anabaena Cylindrica." *Tetrahedron* 60 (2004): 9075–9080.
150. Vilaivan, T.; Winotapan, C.; Banphavichit, V.; Shinada, T.; Ohfune, Y. "Indium-Mediated Asymmetric Barbier-Type Allylation of Aldimines in Alcoholic Solvents: Synthesis of Optically Active Homoallylic Amines." *Journal of Organic Chemistry* 70 (2005): 3464–3471.
151. Friestad, G. K.; Marie, J. C.; Deveau, A. M. "Stereoselective Mn-Mediated Coupling of Functionalized Iodides and Hydrazones: A Synthetic Entry to the Tubulysin γ-Amino Acids." *Organic Letters* 6 (2004): 3249–3252.
152. Planas, L.; Perard-Viret, J.; Royer, J. "Efficient Access to Enantiomerically Pure Rigid Diamines." *Tetrahedron: Asymmetry* 15 (2004): 2399–2403.
153. Ishida, K.; Matsuda, H.; Okita, Y.; Murakami, M. "Aeruginoguanidines 98-A-98-C: Cytotoxic Unusual Peptides from the Cyanobacterium Microcystis Aeruginosa." *Tetrahedron* 58 (2002): 7645–7652.
154. Sharma, G. V. M.; Yadav, T. A.; Choudhary, M.; Kunwar, A. C. "Design of β-Amino Acid With Backbone-Side Chain Interactions: Stabilization of 14/15-Helix in α/β-Peptides." *Journal of Organic Chemistry* 77 (2012): 6834–6848.
155. Palomo, C.; Oiarbide, M.; Landa, A.; Gonzalez-Rego, M. C.; García, J. M.; González, A.; Odriozola, J. M.; Martín-Pastor, M.; Linden, A. "Design and Synthesis of a Novel Class of Sugar-Peptide Hybrids: C-Linked Glyco β-Amino Acids through a Stereoselective 'Acetate' Mannich Reaction as the Key Strategic Element." *Journal of the American Chemical Society* 124 (2002): 8637–8643.
156. Liu, Z.; Bittman, R. "Synthesis of C-Glycoside Analogues of α-Galactosylceramide via Linear Allylic C-H Oxidation and Allyl Cyanate to Isocyanate Rearrangement." *Organic Letters* 14 (2012): 620–623.

157. Noole, A.; Pehk, T.; Järving, I.; Lopp, M.; Kanger, T. "Organocatalytic Asymmetric Synthesis of Trisubstituted Pyrrolidines via a Cascade Reaction." *Tetrahedron: Asymmetry* 23 (2012): 188–198.
158. Schultz-Fademrecht, C.; Kinzel, O.; Marko, I. E.; Pospisil, T.; Pesci, S.; Rowley, M.; Jones, P. "A General Approach to Homochiral α-Amino Substituted Bromo-Heteroaromatics Suitable for Two-dimensional Rapid Analogue Synthesis." *Tetrahedron* 65 (2009): 9487–9493.
159. Kinzel, O. D.; Monteagudo, E.; Muraglia, E.; Orvieto, F.; Pescatore, G.; Ferreira, M.; Rowley, M.; Summa, V. "The Synthesis of Tetrahydropyridopyrimidones as a New Scaffold for HIV-1 Integrase Inhibitors." *Tetrahedron Letters* 48 (2007): 6552–6555.
160. Dondoni, A.; Massi, A.; Sabbatini, S. "Multiple Component Approaches to C-Glycosyl β-Amino Acids by Complementary One-Pot Mannich-Type and Reformatsky-Type Reactions." *Chemistry: A European Journal* 11 (2005): 7110–7125.
161. Smanski, M. J.; Casper, J.; Peterson, R. M.; Yu, Z.; Rajski, S. R.; Shen, B. "Expression of the Platencin Biosynthetic Gene Cluster in Heterologous Hosts Yielding New Platencin Congeners." *Journal of Natural Products* 75 (2012): 2158–2167.
162. Berti, F.; Forzato, C.; Furlan, G.; Nitti, P.; Pitacco, G.; Valentin, E.; Zangrando, E. "Synthesis of Optically Active α-Benzyl Paraconic Acids and their Esters and Assignment of their Absolute Configuration." *Tetrahedron: Asymmetry* 20 (2009): 313–321.
163. Berti, F.; Felluga, F.; Forzato, C.; Furlan, G.; Nitti, P.; Pitacco, G.; Valentin, E. "Chemoenzymatic Synthesis of Diastereomeric Ethyl γ-Benzyl Paraconates and Determination of the Absolute Configurations of their Acids." *Tetrahedron: Asymmetry* 17 (2006): 2344–2353.
164. Ammazzalorso, A.; Bettoni, G.; De Filippis, B.; Fantacuzzi, M.; Giampietro, L.; Giancristofaro, A.; Maccallini, C.; Re, N.; Amoroso, R.; Coletti, C. "Synthesis of 2-Aryloxypropanoic Acids Analogues of Clofibric Acid and Assignment of the Absolute Configuration by $^1$H NMR Spectroscopy and DFT Calculations." *Tetrahedron: Asymmetry* 19 (2008): 989–997.
165. Seco, J. M.; Quiñoá, E.; Riguera, R "Simplified NMR Procedures for the Assignment of the Absolute Configuration." In *Structure Elucidation in Organic Chemistry: The Search for the Right Tools*, ed. María Magdalena Cid and Jorge Bravo. Weinheim, Germany: Wiley-VCH Verlag GmbH & Co. KGaA, 2015.
166. Ciminiello, P.; Dell'Aversano, C.; Fattorusso, E.; Forino, M.; Magno, S.; Di Rosa. M.; Ianaro. A.; Poletti, R. "Structure and Stereochemistry of a New Cytotoxic Polychlorinated Sulfolipid from Adriatic Shellfish." *Journal of the American Chemical Society* 124 (2002): 13114–13120.
167. Nilewski, C.; Deprez, N. R.; Fessard, T. C.; Li, D. Bo; Geisser, R. W.; Carreira, E. M. "Synthesis of Undecachlorosulfolipid A: Re-evaluation of the Nominal Structure." *Angewandte Chemie, International Edition* 50 (2011): 7940–7943.
168. Ondeyka, J.; Buevich, A. V.; Williamson, R. T.; Zink, D. L.; Polishook, J. D.; Occi, J.; Vicente, F.; Basilio, A.; Bills, G. F.; Donald, R. G. K.; Phillips,, J. W.; Goetz,, M. A.; Singh, S. B. "Isolation, Structure Elucidation, and Biological Activity of Altersolanol P Using Staphylococcus Aureus Fitness Test Based Genome-Wide Screening." *Journal of Natural Products* 77 (2014): 497–502.

169. Arens, J C.; Berrue, F; Pearson, J. K.; Kerr, R. G. "Isolation and Structure Elucidation of Satosporin A and B: New Polyketides from Kitasatospora Griseola." *Organic Letters* 15 (2013): 3864–3867.
170. Mancilla, G.; Femenía-Ríos, M.; Grande, M.; Hernández-Galán, R.; Macías-Sánchez, A. J.; Collado, I. G. "Enantioselective, Chemoenzymatic Synthesis, and Absolute Configuration of the Antioxidant (−)-Gloeosporiol." *Tetrahedron* 66 (2010): 8068–8075.
171. Gross, H.; McPhail, K. L.; Goeger, D. E..; Valeriote, F. A.; Gerwick, W. H. "Two Cytotoxic Stereoisomers of Malyngamide C, 8-Epi-Malyngamide C and 8-O-Acetyl-8-Epi-Malyngamide C, from the Marine Cyanobacterium Lyngbya Majuscula." *Phytochemistry* 71 (2010): 1729–1735.
172. Jimenez-Teja, D.; Daoubi, M.; Collado, I. G.; Hernández-Galán, R. "Lipase-Catalyzed Resolution of 5-Acetoxy-1,2-Dihydroxy-1,2,3,4-Tetrahydronaphthalene. Application to the Synthesis of (+)-(3R,4S)-Cis-4-Hydroxy-6-Deoxyscytalone, a Metabolite Isolated from Colletotrichum Acutatum." *Tetrahedron* 65 (2009): 3392–3396.
173. Lira, S. P.; Vita-Marques, A. M.; Seleghim, M. H. R.; Bugni, T. S.; LaBarbera, D. V.; Sette, L. D.; Sponchiado, S. R. P.; Ireland, C. M.; Berlinck, R. G. S. "New Destruxins from the Marine-Derived Fungus Beauveria Felina." *Journal of Antibiotics* 59 (2006): 553–563.
174. Perez-García, E.; Zubía, E.; Ortega, M. J.; Carballo, J. L. "Merosesquiterpenes from Two Sponges of the Genus Dysidea." *Journal of Natural Products* 68 (2005): 653–658.
175. Williamson, R. T.; Boulanger, A.; Vulpanovici, A.; Roberts, M. A.; Gerwick, W. H. "Structure and Absolute Stereochemistry of Phormidolide, a New Toxic Metabolite from the Marine Cyanobacterium Phormidium Sp." *Journal of Organic Chemistry* 67 (2002): 7927–7936.
176. Reyes, F.; Arda, A.; Martín, R.; Fernández, R.; Rueda, A.; Montalvo, D.; Gómez, C.; Jimenez, C.; Rodriguez, J.; Sanchez-Puelles, J. M. "New Cytotoxic Cembranes from the Sea Pen Gyrophyllum Sibogae." *Journal of Natural Products* 67 (2004): 1190–1192.
177. Cohen, E.; Koch, L.; Thu, K. M.; Rahamim, Y.; Aluma, Y.; Ilan, M.; Yarden, O.; Carmeli, S. "Novel Terpenoids of the Fungus Aspergillus Insuetus Isolated from the Mediterranean Sponge Psammocinia Sp. Collected Along the Coast of Israel." *Bioorganic and Medicinal Chemistry* 19 (2011): 6587–6593.
178. Cedron, J. C.; Oberti, J. C.; Estevez-Braun, A.; Ravelo, A. G.; Del Arco-Aguilar, M.; Lopez, M. "Pancratium Canariense as an Important Source of Amaryllidaceae Alkaloids." *Journal of Natural Products* 72 (2009): 112–116.
179. Habib, E.; Leon, F.; Bauer, J. D.; Hill, R. A.; Carvalho, P.; Cutler, H. G.; Cutler, S. J. "Mycophenolic Derivatives from Eupenicillium Parvum." *Journal of Natural Products* 71 (2008): 1915–1918.
180. Rojas-Cabrera, H.; Fernández-Zertuche, M.; García-Barradas, O.; Muñoz-Muñiz, O.; Ordóñez, M. "Preparation of Dimethyl (R)- and (S)-2-(2-Hydroxyphenyl)-2-Hydroxyethylphosphonate Derived from Salicylaldehyde via Resolution Using (S)-Methoxyphenylacetic Acid (MPA)." *Tetrahedron: Asymmetry* 18 (2007): 142–148.
181. Gesner, S.; Cohen, N.; Ilan, M.; Yarden, O.; Carmeli, S. "Pandangolide 1a, a Metabolite of the Sponge-Associated Fungus Cladosporium Sp., and the Absolute Stereochemistry of Pandangolide 1 and Iso-cladospolide B." *Journal of Natural Products* 68 (2005): 1350–1353.

182. Klemke, C.; Kehraus, S.; Wright, A. D.; Koenig, G. M. "New Secondary Metabolites from the Marine Endophytic Fungus Apiospora Montagnei." *Journal of Natural Products* 67 (2004): 1058–1063.
183. Bosque, I.; Foubelo, F.; Gonzalez-Gomez, J. C. "A General Protocol to Afford Enantioenriched Linear Homoprenylic Amines." *Organic and Biomolecular Chemistry* 11 (2013): 7507–7515.
184. Remuinan, M. J.; Pattenden, G. "Total Synthesis of (-)-Pateamine, a Novel Polyene Bis-Macrolide with Immunosuppressive Activity from the Sponge Mycale Sp." *Tetrahedron Letters* 41 (2000): 7367–7371.
185. Earle, M. A.; Hultin, P. G. "4-Oxo α-Amino Acids: A Caution When Determining Absolute Configurations by $^1$H NMR Using MPA Derivatization and Ba$^{2+}$ Complexation." *Tetrahedron Letters* 41 (2000): 7855–7858.
186. Abad, J. L.; Camps, F. "Arylacetic Acid Derivatization of 2,3- and Internal Erythro-Squalene Diols. Separation and Absolute Configuration Determination." *Tetrahedron* 60 (2004): 11519–11525.
187. Yang, S.; Shen, Tao; Lijuan, Z.; Li, C.; Zhang, Y.; Lou, H.; Ren, D. "Chemical Constituents of Lobelia Chinensis." *Fitoterapia* 93 (2014): 168–174.
188. Santos, S.; Cabral, V.; Graca, J. "Cork Suberin Molecular Structure: Stereochemistry of the C18 Epoxy and Vic-Diol ω-Hydroxyacids and α,ω-Diacids Analyzed by NMR." *Journal of Agricultural and Food Chemistry* 61 (2013): 7038–7047.
189. O'Neill, T.; Johnson, J. A.; Webster, D.; Gray, C. A. "The Canadian Medicinal Plant Heracleum Maximum Contains Antimycobacterial Diynes and Furanocoumarins." *Journal of Ethnopharmacology* 147 (2013): 232–237.
190. Guo, J. P.; Zhu, C. Y.; Zhang,. P.; Chu, Y. S.; Wang, Y. L.; Zhang, J. X.; Wu, D. K.; Zhang, K. Q.; Niu, X. M. "Thermolides, Potent Nematocidal PKS-NRPS Hybrid Metabolites from Thermophilic Fungus Talaromyces Thermophilus." *Journal of the American Chemical Society* 134 (2012): 20306–20309.
191. Singh, A. A.; Zulkifli, S. N. A.; Meyns, M.; Hayes, P. Y.; De Voss, J. J. "Synthesis of Highly Enantioenriched Hydroxy- and Dihydroxy-Fatty Esters: Substrate Precursors For Cytochrome P 450Biol." *Tetrahedron: Asymmetry* 22 (2011): 1709–1719.
192. Takamura, H.; Kadonaga, Y.; Kadota, I.; Uemura, D. "Stereo-Controlled Synthesis and Structural Confirmation of the C14-C24 Degraded Fragment of Symbiodinolide." *Tetrahedron* 66 (2010): 7569–7576.
193. Han, C.; Yamano, Y.; Kita, M.; Takamura, H.; Uemura, D. "Determination of Absolute Configuration of C14-C23 Fragment in Symbiodinolide." *Tetrahedron Letters* 50 (2009): 5280–5282.
194. Murata, T.; Sano, M.; Takamura, H.; Kadota, I.; Uemura, D. "Synthesis and Structural Revision of Symbiodinolide C23-C34 Fragment." *Journal of Organic Chemistry* 74 (2009): 4797–4803.
195. Takamura, H.; Kadonaga, Y.; Yamano, Y.; Han, C.; Kadota, I.; Uemura, D. "Stereoselective Synthesis and Absolute Configuration of the C33-C42 Fragment of Symbiodinolide." *Tetrahedron* 65 (2009): 7449–7456.
196. Takamura, H.; Kadonaga, Y.; Yamano, Y.; Han, C.; Aoyama, Y.; Kadota, I.; Uemura, D. "Synthesis and Structural Determination of the C33-C42 Fragment of Symbiodinolide." *Tetrahedron Letters* 50 (2009): 863–866.

197. Oh, D. C.; Scott, J. J.; Currie, C. R.; Clardy, J. "Mycangimycin, a Polyene Peroxide from a Mutualist Streptomyces Sp." *Organic Letters* 11 (2009): 633–636.
198. Caballero, W.; Shindo, S.; Murakami, T.; Hashimoto, M.; Tanaka, K.; Takada, N. "Absolute Stereochemistry and Conformational Analysis of Achaetolide Isolated from Ophiobolus Sp." *Tetrahedron* 65 (2009): 7464–7467.
199. Kwon, H. C.; Kauffman, C. A.; Jensen, P. R.; Fenical, W. "Marinisporolides, Polyene-Polyol Macrolides from a Marine Actinomycete of the New Genus Marinispora." *Journal of Organic Chemistry* 74 (2009): 675–684.
200. Phuwapraisirisan, P.; Phoopichayanun, C.; Supudompol, B. "Feroniellic Acids A–C, Three New Isomeric Furanocoumarins with Highly Hydroxylated Geranyl Derived Moieties from Feroniella Lucida." *Tetrahedron Letters* 49 (2008): 3133–3136.
201. Oguchi, K.; Tsuda, M.; Iwamoto, R.; Okamoto, Y.; Kobayashi, J.; Fukushi, E.; Kawabata, J.; Ozawa, T.; Masuda, A.; Kitaya, Y.; Omasa, K. "Iriomoteolide-3a, a Cytotoxic 15-Membered Macrolide from a Marine Dinoflagellate Amphidinium Species." *Journal of Organic Chemistry* 73 (2008): 1567–1570.
202. Williams, P. G.; Miller, E. D.; Asolkar, R. N.; Jensen, P. R.; Fenical, W. "Arenicolides A–C, 26-Membered Ring Macrolides from the Marine ActinomyceteSalinispora Arenicola." *Journal of Organic Chemistry* 72 (2007): 5025–5034.
203. Kwon, H. C.; Kauffman, C. A.; Jensen, P. R.; Fenical, W. "Marinomycins A–D, Antitumor-Antibiotics of a New Structure Class from a Marine Actinomycete of the Recently Discovered Genus 'Marinispora.'" *Journal of the American Chemical Society* 128 (2006): 1622–1632.
204. Tsuda, M.; Izui, N.; Shimbo, K.; Sato, M.; Fukushi, E.; Kawabata, J.; Katsumata, K.; Horiguchi, T.; Kobayashi, J. "Amphidinolide X, a Novel 16-Membered Macrodiolide from Dinoflagellate Amphidinium Sp." *Journal of Organic Chemistry* 68 (2003): 5339–5345.
205. Duret, P.; Waetcher, A.; Figadere, B.; Hocquemiller, R.; Cavé, A. "Determination of Absolute Configurations of Carbinols of Annonaceous Acetogenins with 2-Naphthylmethoxyacetic Acid Esters." *Journal of Organic Chemistry* 63 (1998): 4717–4720.
206. Sinz, A.; Matusch, R.; Kämpchen, T.; Fiedler, W.; Schmidt, J.; Santisuk, T.; Wangcharoentrakul, S.; Chaichana, S.; Reutrakul, V. "Novel Acetogenins from the Leaves of Dasymaschalon Sootepense." *Helvetica Chimica Acta* 81 (1998): 1608–1615.
207. Craig, D.; Alali, F. Q.; Gu, Z.; McLaughlin, J. L. "Three New Bioactive Bis-Adjacent THF-ring Acetogenins from the Bark of Annona Squamosa." *Bioorganic and Medicinal Chemistry* 6 (1998): 569–575.
208. Craig, D.; Zeng, L.; Gu, Z.; Kozlowski, J. F.; McLaughlin, J. L. "Novel Mono-Tetrahydrofuran Ring Acetogenins, from the Bark of Annona Squamosa, Showing Cytotoxic Selectivities for the Human Pancreatic Carcinoma Cell Line, PACA-2." *Journal of Natural Products* 60 (1997): 581–586.
209. He, K.; Zhao, G.; Shi, G.; Zeng, L.; Chao, J.; McLaughlin, J. L. "Additional Bioactive Annonaceous Acetogenins from Asimina Triloba (Annonaceae)." *Bioorganic and Medicinal Chemistry* 5 (1997): 501–506.
210. Jiang, Z.; Yu, D. Q. "New Type of Mono-Tetrahydrofuran Ring Acetogenins from Goniothalamus Donnaiensis." *Journal of Natural Products* 60 (1997): 122–125.

211. Woo, M.; Yol, K.; Cho, Y.; Zhang, Y.; Zeng, L.; Gu, Z.; McLaughlin, J. L. "Asimilobin and Cis- and Trans-Murisolinones, Novel Bioactive Annonaceous Acetogenins from the Seeds of Asimina Triloba." *Journal of Natural Products* 58 (1995): 1533–1542.
212. Shimada, H.; Nishioka, S.; S.; Singh, S.; Sahai, M.; Fujimoto, Y. "Absolute Stereochemistry of Non-Adjacent Bis-Tetrahydrofuranic Acetogenins." *Tetrahedron Letters* 35 (1994): 3961–3964.
213. Rodríguez, J.; Riguera, R.; Débitus, C. "The Natural Polypropionate-Derived Esters of the Mollusk Onchidium Sp." *Journal of Organic Chemistry* 57 (1992): 4624–4632.
214. Ciavatta, M. L.; Manzo, E.; Nuzzo, G.; Villani, G.; Varcamonti, M.; Gavagnin, M. "Crucigasterins A–E, Antimicrobial Amino Alcohols from the Mediterranean Colonial Ascidian Pseudodistoma Crucigaster." *Tetrahedron* 66 (2010): 7533-7538.
215. Uekoa, R.; Fujita, T.; Iwashita, T.; Soest, R. W. M.; Matsunaga, S. "Inconspicamide, New N-Acylated Serinol from the Marine Sponge Stelletta Inconspicua." *Bioscience, Biotechnology, and Biochemistry* 72 (2008): 3055–3058.
216. Kobayashi, J.; Tsuda, M.; Cheng, J.; Ishibashi, M.; Takikawa, H.; Mori, K. "Absolute Stereochemistry of Penaresidins A and B." *Tetrahedron Letters* 37 (1996): 6775–6776.
217. Prosperini, S.; Pastori, N.; Ghilardi, A.; Clerici, A.; Punta, C. "New Domino Radical Synthesis of Aminoalcohols Promoted by TiCl4-Zn/t-BuOOH System: Selective Hydroxyalkylation of Amines in Alcohol or in Cyclic Ether Cosolvents." *Organic and Biomolecular Chemistry* 9 (2011): 3759–3767.
218. Clerici, A.; Ghilardi, A.; Pastori, N.; Punta, C.; Porta, O. "A New One-Pot, Four-Component Synthesis of 1,2-Amino Alcohols: TiCl3/t-BuOOH-Mediated Radical Hydroxymethylation of Imines." *Organic Letters* 10 (2008): 5063–5066.
219. Kobayashi, H.; Ohashi, J.; Fujita, T.; Iwashita, T.; Nakao, Y.; Matsunaga, S.; Fusetani, N. "Complete Structure Elucidation of Shishididemniols: Complex Lipids With Tyramine-Derived Tether and Two Serinol Units, from a Marine Tunicate of the Family Didemnidae." *Journal of Organic Chemistry* 72 (2007): 1218–1225.
220. Kobayashi, H.; Miyata, Y.; Okada, K.; Fujita, T.; Iwashita, T.; Nakao, Y.; Fusetania, N.; Matsunagaa, S. "The Structures of Three New Shishididemniols from a Tunicate of the Family Didemnidae." *Tetrahedron* 63 (2007): 6748–6754.

# INDEX

*ab initio*, 6
(S)-3-(acetylthio)-2-methylpropanoic acid
  9-AHA esters, assignment,
    $^1$H NMR, 87–89
9-AHA
  carboxylic acids, 87–91
  conformation, 88
  preparation, 28
  preparation of derivatives, 28–34
  structure, 4
aldehyde cyanohydrins
  conformational composition, main
    conformers, MPA derivatives, 59
  correlation model, 58–59
  selected CDA and NMR conditions, 58
  shielding effects, 58
  validation structures, 63
9-AMA
  *prim/sec*-1,2-diols, 140
  *sec/sec*-diols, 112–120
  esterification shifts, 16, 105–110
  preparation, 28
  preparation of derivatives, 28–35
  primary alcohols, 51–57
  secondary alcohols, 37–46
  shielding effect, 10–11
  structure, 4
*prim/sec*-1,2-amino alcohols
  $\Delta\delta^R$, $\Delta\delta^S$, 174–175, 182
  $\Delta\delta^{RS}$, OMe and C$\alpha$H, 174–175
  $\Delta\delta^{RS}$, R and methylene, 174
  $\Delta\delta^{T1T2}$, 183–184
  conformational composition, main
    conformers, MPA derivatives, 173–174
  correlation model, 172, 174, 182
  cross interaction, 174
  diagnostic signals, 173–174
  double derivatization, 173–181
  selected CDA and NMR conditions, 172, 182
  shielding effects, 174, 183–184
  single derivatization, 182–187
  validation structures, 188

*sec/prim*-1,2-amino alcohols
  $\Delta\delta^R$, $\Delta\delta^S$, 158–159, 166
  $\Delta\delta^{RS}$, R and methylene, 156
  $\Delta\delta^{RS}$, OMe and C$\alpha$H, 158
  $\Delta\delta^{T1T2}$, 165
  conformational composition, main
    conformers, MPA derivatives, 156
  correlation model, 155, 156, 160, 166
  cross interaction, 157
  diagnostic signals, 155, 166
  double derivatization, 155–164
  selected CDA and NMR conditions, 155, 166
  shielding effects, 156, 160, 167, 168
  single derivatization, 165–171
  validation structures, 171
*sec/sec*-1,2-amino alcohols
  $\Delta\delta^{RS}$ sign distribution, 130
  conformational composition, main
    conformers, MPA derivatives, 128
  correlation model, 129–130
  diagnostic signals, 129–130
  selected CDA and NMR conditions, 128
  shielding effects, 129
  structural types, 128
  validation structures, 134
2-amino-3-methylpentan-1-ol
  *bis*-MPA derivatives, exercise, 217
2-amino-4-methylpentan-1-ol
  *bis*-(S)-MPA derivative, low temperature, exercise, 219
2-amino-4-(methylthio)butan-1-ol
  *bis*-MPA amido esters, exercise, 216, 217
(R)-2-amino-3-methylbutan-1-ol
  *bis*-MPA amidoesters, assignment,
    $^1$H NMR, shielding effects, sign distribution, 163–165
1-amino-4-phenylbutan-2-ol
  *bis*-MPA derivatives, exercise, 219
2-aminobutan-1-ol
  *bis*-(S)-MPA amido ester, exercise, 187

1-aminoheptan-2-ol
  bis-(R)-MPA amido ester, exercise, 221
(R)-1-aminoheptan-2-ol
  bis-MPA amidoesters, assignment,
    $^1$H NMR, shielding effects, sign
    distribution 179–181
2-aminopentan-3-ol
  bis-MPA amidoesters, assignment,
    $^1$H NMR, 130–132
  bis-MPA amidoesters, exercise, 213
(R)-1-aminopropan-2-ol
  bis-(R)-MPA amidoesters, assignment
    by low temperature, $^1$H NMR,
    shielding effects, sign distribution,
    186–187
(S)-1-aminopropan-2-ol
  bis-MPA amidoesters, assignment by
    double derivatization, $^1$H NMR,
    shielding effects, sign distribution,
    175–179
  bis-(R)-MPA amidoesters, assignment by
    low temperature, $^1$H NMR, shielding
    effects, sign distribution, 184
(R)-2-aminopropan-1-ol
  bis-(R)-MPA amidoesters, assignment by
    low temperature, $^1$H NMR, shielding
    effects, sign distribution, 168
(S)-2-aminopropan-1-ol
  bis-MPA amidoesters, assignment,
    double derivatization, $^1$H NMR,
    157, 159
  bis-(R)-MPA amidoesters, assignment by
    low temperature, $^1$H NMR, 167
ap conformer
  9-AHA esters, 88
  9-AMA esters, secondary
    alcohols, 6, 17
  9-AMA esters, primary alcohols, 53
  BPG amides, 77
  MPA amides, 15, 81
  MPA esters, 6–7, 14, 15, 38
  MPA thioesters, 69
  MPA, aldehyde cyanhohydrins, 58
  MPA, ketone cyanhohydrins, 65
  2-NTBA thioesters, 69
ap1 conformer
  MTPA amides, 84
  MTPA esters, 48
ap3 conformer
  MTPA amides, 84

ap-sp-gg-I conformer, 175
ap-sp-gt-II conformer, 175

borneol
  MPA esters, exercise, 198
  MTPA esters, (−)-borneol, assignment,
    $^1$H NMR, 47
bornylamine
  BPG amides, exercise, 202
  MPA amides, (−)-bornylamine,
    assignment, $^1$H NMR, 82
  MTPA amides, (−)-bornylamine,
    assignment, $^1$H NMR, 83
BPG
  amides, main conformers, 77
  preparation of derivatives, 28–31
  primary amines, 76–80
  structure, 4
butan-2-amine
  MPA, exercise, 203
(S)-butan-2-amine
  BPG amides, assignment, $^{13}$C NMR, 79
butan-2-ol
  (R)-MPA ester, low temperature,
    exercise, 205
(R)-butan-2-ol
  MPA esters, assignment, $^{13}$C NMR, 42
  (S)-MPA ester, low temperature,
    assignment, $^1$H NMR, 98
(S)-butan-2-ol
  mix-and-shake, 33
  1:2 mixture of (R)- and (S)-MPA,
    assignment, $^1$H NMR, 44
(S)-butane-2-thiol
  MPA thioesters, assignment, $^1$H NMR, 71

carbonyl effects
  9-AMA esters of prim/sec-1,2-diols,
    140–142, 145
  MPA esters of prim/sec-1,2-diols,
    136–137, 139
α-chiral carboxylic acids
  conformational composition, main
    conformers, 9-AHA, 88–89
  correlation model, 9-AHA, 88
  exercises, 204
  selected CDA and NMR conditions, 87
  shielding effects, 82
  validation structures, 90
chiral derivatizing agents (CDA), 1, 3–4

β-chiral primary alcohols
    conformational composition of 9-AMA esters, 53
    correlation model, 53
    polar L1/L2 groups, 53
    selected CDA and NMR conditions, 52
    shielding effects, 53
    validation structures, 57
α-chiral primary amines
    BPG, correlation model, main conformers, $\Delta\delta^{RS}$, 76–77
    complexation with $Ba^{2+}$, $\Delta\delta^{Ba}$, 94
    double derivatization, 76
    MPA, correlation model, main conformers, $\Delta\delta^{RS}$, 80
    MTPA, correlation model, main conformers, $\Delta\delta^{SR}$, 83
    selected CDA and NMR conditions, 77
    single derivatization, 102
    validation structures, 86
chiral solvating agent (CSA), 1
chiral thiols
    MPA, correlation model, main conformers, $\Delta\delta^{RS}$, 70
    2-NTBA, correlation model, main conformers, $\Delta\delta^{RS}$, 70
    selected CDA and NMR conditions, 70
    validation structures, 75
1,4-*bis*-O-(4-chlorobenzyloxy)-D-threitol (*syn*)
    *bis*-MPA ester, assignment, $^{13}$C NMR, 119
(S)-2-chloropropan-1-ol
    9-AMA esters, assignment, $^1$H NMR, 55
complexation with $Ba^{2+}$
    $\Delta\delta^{Ba}$, 14–15
    primary amines, 14–15, 102–105
    secondary alcohols, 14–15, 99–102
correlation models, 21–22
criteria for correct assignment, 18–21
cross interaction
    *bis*-MPA derivatives of *prim/sec*-1,2-amino alcohols, 174–175
    *bis*-MPA derivatives of *sec/prim*-1,2-amino alcohols, 157–158

derivatization
    acid chlorides, 29
    amines, 28
    carboxylic acids, 28

CDA-resin, 30–31
cyanohydrins, 28
*prim/sec*-1,2-diols, 28
*sec/sec*-1,2-diols, 28
in tube, amines, 32
in tube, aminoalcohols, 34
in tube, diols, 35
in tube, triols, 35
in tube, alcohols, cyanohydrins, and thiols, 32
primary alcohols, 29
secondary alcohols, 29
thiols, 29
*prim/sec/sec*-1,2,3-triols, 28
diacetone D-glucose
    low temperature procedure using (R)-MPA, 95
    MPA esters, assignment, $^1$H NMR, 39
diacetone glucose
    (R)-MPA ester, $Ba(ClO_4)_2$, exercise, 207
diagnostic signals
    monofunctional substrates, 24
    polyfunctional substrates, 23
3,3-difluoroheptane-1,2-diol
    (R)-MPA, low temperature, exercise, 214
3,4-dihydroxy-5-methylhexan-2-one
    low temperature procedure using (R)-MPA, 124–125
2,3-dihydroxyhexyl acetate
    MPA esters, exercise, 209
(R)-2,3-dihydroxypropyl stearate
    9-AMA esters, assignment, $^1$H NMR, 144
3,3-dimethylbutan-2-ol
    (R)-9-AMA esters, chiral HPLC, 208
*prim/sec*-1,2-diols
    $\Delta\delta^{RS}$, 9-AMA, 135, 141
    $\Delta\delta^{R}$, $\Delta\delta^{S}$, 9-AMA, 135, 142
    $\Delta\delta^{RS}$, MPA, 137
    $\Delta\delta^{T1T2}$, 148, 149, 150
    conformational composition, main conformers, 9-AMA, 136, 141, 142
    conformational composition, main conformers, MPA, 136, 137
    diagnostic signals, 121, 132
    double derivatization, 135
    selected CDA and NMR conditions, 135, 136
    shielding effects, 137, 138, 140, 141, 142
    single derivatization, 147
    validation structures, 154

sec/sec-1,2 and 1,n-diols
  $\Delta\delta^{RS}$, 113
  $\Delta\delta^{T1T2}$, 122–124
  conformational composition, main conformers, MPA and 9-AMA derivatives, 112–124
  correlation model, 115
  diagnostic signals, 114, 115
  double derivatization, 110–115
  selected CDA and NMR conditions, 110
  single derivatization, 121–124
  shielding effects, 113, 114, 123
  structural types, 112
  validation structures, 127
double derivatization methods
  aldehyde cyanohydrins, 58
  prim/sec-1,2-amino alcohols, 152
  sec/prim-1,2-amino alcohols, 146
  sec/sec-1,2-amino alcohols, 155
  carboxylic acids, 87
  prim/sec-1,2-diols, 135
  sec/sec-diols, 111
  ketone cyanohydrins, 63
  primary alcohols, 51
  primary amines, 76
  secondary alcohols, 37
  prim/sec/sec-1,2,3-triols, 188
  thiols, 69

esterification shifts, 16, 105–110
ethyl 2-mercaptopropanoate
  MPA thioesters, exercise, 202
(R)-ethyl 2-mercaptopropanoate
  2-NTBA thioesters, assignment, $^1$H NMR, 72
  2-NTBA thioesters, assignment, $^{13}$C NMR, 73
exercises
  prim/sec-1,2-amino alcohols, double, 219–220
  prim/sec-1,2-amino alcohols, single, 220–221
  sec/prim-1,2-amino alcohols, double, 216–218
  sec/prim-1,2-amino alcohols, single, 218–219
  sec/sec-1,2-amino alcohols, 213
  carboxylic acids, 204
  sec/sec-diols, double, 209–211
  sec/sec-diols, single, 211–212

prim/sec-1,2-diols, double, 215–216
prim/sec-1,2-diols, single, 214–215
ketone cyanohydrins, 200–201
primary alcohols, 199–200
primary amines, double, 202–204
primary amines, single, 207–208
secondary alcohols, double, 197–198
secondary alcohols, single, complexation, 206–207
secondary alcohols, single, esterification shifts, 208–209
secondary alcohols, single, low temperature, 205, 181
prim/sec/sec-1,2,3-triols, 221–222
thiols, 201–202

heptane -2,3-diol
  bis-MPA esters, heptane-2,3-diol (anti), assignment $^1$H NMR, 118
  bis-MPA esters, heptane-2,3-diol (syn), assignment $^1$H NMR, 116
  bis-(R)-MPA esters, heptane-2,3-diol, low temperature, exercise, 212
hexane -1,2,3-triol (anti)
  tris-MPA derivatives, assignment, $^1$H NMR, 193
hexane -1,2,3-triol (syn)
  tris-MPA derivatives, assignment, $^1$H NMR, 192
(S)-2-hydroxy-2,4-dimethylpentanenitrile
  MPA esters, assignment, $^{13}$C NMR, 67
(1R, 2S, 5R)-1-hydroxy-2-isopropyl-5-methylcyclohexanecarbonitrile
  MPA esters, assignment, $^1$H NMR, 65
(R)-2-hydroxy-2-(4-methoxyphenyl)acetonitrile
  MPA esters, assignment, $^{13}$C NMR, 60
(R)-2-hydroxy-3-methylbutanenitrile
  MPA esters, assignment, $^1$H NMR, 59
(1R, 4S)-hydroxyciclopent-2-en-1-yl acetate
  9-AMA, esterification shifts, $^1$H NMR, 107

isopinocampheol
  9-AMA esters, exercise, 197
(−)-isopinocampheylamine
  BPG amides, assignment, $^1$H NMR, 78
  MPA amides, complexation with $Ba^{2+}$, exercise, 207

MPA amides, complexation with $Ba^{2+}$, $^1H$ NMR, 105
1,2-isopropylidene-glycerol
  9-AMA esters, exercise, 199
(−)-isopulegol
  9-AMA, assignment, $^1H$ NMR, 40

ketone cyanohydrins
  conformational composition, main conformers, MPA derivatives, 64
  correlation model, 64
  selected CDA and NMR conditions, 64
  shielding effects, 64
  validation structures, 68

leucine methyl ester
  BPG amides, exercise, 204
loading of the CDA-resins, 31
low temperature NMR, 13, 93

menthol
  1:3 mixture of MPA, exercise, 199
  aromatic shielding effects win different AMAAs, 11
  (−)-menthol, assignment, 1:2 mixture of (R)- and (S)-9-AMA, 45
  MPA, complexation with $Ba^{2+}$, exercise, 206
methine protons
  Hα(R1), Hα(R2), 111
methyl 4-amino-3-hydroxy-5-phenylpentanoate (anti)
  bis-MPA amido esters, assignment using $^1H$ NMR, 132
methyl 3-hydroxy-2-methylpropanoate
  9-AMA esters, exercise, 200
(S)-2-methylbutan-1-ol
  9-AMA esters, assignment, $^1H$ NMR, 52
2-methylbutanoic acid
  9-AHA esters, exercise, 204
mix-and-shake procedure, 14, 18, 29
molecular mechanics, 6
monofunctional substrates, 24
MPA
  aldehyde cyanohydrins, 58
  prim/sec-1,2-amino alcohols, 172, 182
  sec/prim-1,2-amino alcohols, 155, 166
  sec/sec-1,2-amino alcohols, 128
  prim/sec-1,2-diols, 135, 148
  sec/sec-diols, 110, 121
  ketone cyanohydrins, 64

preparation of derivatives, 27–35
primary amines, 77, 80, 102
secondary alcohols, 38, 95, 99
structure, 4
thiols, 69
prim/sec/sec-1,2,3-triols, 189
MTPA
  sec/sec-diols, 100
  preparation of derivatives, 27–35
  primary amines, 77, 83
  secondary alcohols, 38, 47
  structure, 4
trans-muriosolinone
  bis-MTPA esters, exercise, 210

1-NMA
  preparation, 27
  preparation of derivatives, 27–35
  sec/sec-diols, 111
  secondary alcohols, 38
  structure, 4
2-NMA
  sec/sec-diols, 111
  preparation, 27
  preparation of derivatives, 27–35
  secondary alcohols, 38
  structure, 4
NMR parameters
  $\Delta\delta^{Ba}$, 15, 16
  $\Delta\delta^{AR}, \Delta\delta^{AS}$, 105, 106
  $\Delta\delta^{RS}$, 2
  $\Delta\delta^{SR}$, 47
  $\Delta\delta^R, \Delta\delta^S$, prim/sec-1,2-diols, 142
  $\Delta\delta^R, \Delta\delta^S$, prim/sec and sec/prim-1,2-amino alcohols, 158
  $\Delta\delta^{T1T2}$, 14
  $[|\Delta\delta^{RS}|]$, 191
2-NTBA
  preparation, 27
  preparation of derivatives, 27–35
  structure, 4
  thiols, 69

pentan-2-ol
  MPA esters, general procedure, 9–10
(R)-pentan-2-ol
  (R)-MPA, complexation with $Ba^{2+}$, assignment, $^1H$ NMR, 101
  (S)-MPA, complexation with $Ba^{2+}$, assignment, $^1H$ NMR, 100

1-phenylethane-1,2-diol
  bis-MPA esters, low temperature, exercise, 214
(R)-phenylethane-1,2-diol
  bis-9-AMA esters, assignment, $^1$H NMR, 145
polyfunctional substrates, 23
preparation of
  acid chloride resin, 30
  CDA, 27
  CDA acid chlorides, 29
  CDA derivatives, 27–35
  CDA-resins, 31
(R)-propane-1,2-diol
  bis-(R)-MPA esters, assignment, $^1$H NMR, low temperature, 152
(S)-propane-1,2-diol
  bis-9-AMA esters, double derivatization, assignment, $^1$H NMR, 143
  bis-MPA esters, double derivatization, assignment, $^1$H NMR, 138
  bis-(R)-MPA esters, low temperature, assignment, $^1$H NMR, 150

resin loading, 31
resin scavengers, 32
rolliniastatin-2
  2-NMA ester, exercise, 121

semiempirical, 6
simultaneous derivatization
  (R)/(S)-9-AMA esters, $^{13}$C NMR, 45
  (R)/(S)-MPA amides, exercise, 203
  (R)/(S)-MPA esters, 43–45
  polyfunctional compounds, 111, 127, 134, 155, 188
single derivatization methods
  prim/sec-1,2-aminoalcohols, 181
  sec/prim-1,2-aminoalcohols, 165
  prim/sec-1,2-diols, 147
  sec/sec-1,2-diols, 120
  primary amines, complexation with $Ba^{2+}$, 14, 102
  secondary alcohols, complexation with $Ba^{2+}$, 14, 99

secondary alcohols, esterification shifts, 16, 105
secondary alcohols, low temperature, 13, 93
sp conformer
  9-AHA esters, 88
  9-AMA esters, primary alcohols, 53
  9-AMA esters, secondary alcohols, 6, 14, 17, 28, 42, 106
  BPG amides, 77
  MPA amides, 15, 81
  MPA esters, 6, 14, 15, 38
  MPA thioesters, 70
  MPA, aldehyde cyanohydrins, 59
  MPA, ketone cyanohydrins, 64
  MTPA, amides, 84
  2-NTBA thioesters, 70
sp1 conformer
  MTPA esters, 48
sp2 conformer
  MTPA esters, 48
sp-ap-gt-I conformer, 156, 158
sp-gt-I conformer, 136, 148
sp-gt-II conformer, 136, 148

1-(trimethylsilyl)propane-1,2,3-triol
  tris-MPA esters, exercise, 221
prim/sec/sec-1,2,3-triols NMR
  correlation model, tris-MPA esters, 191
  diagnostic signals, 190, 191
  parameters for assignment, $\Delta\delta^{RS}H(3')$, $\Delta\delta^{RS}H(2')$, $|\Delta(\Delta\delta^{RS})|$, 191
  selected CDA and NMR conditions, 189
  structural types, 189
  shielding effects, 190
  validation structures, 194
in tube derivatization of
  alcohols, cyanohydrins, and thiols, 32
  amines, 31
  aminoalcohols, 34
  diols, 35
  triols, 35